NAVIGATING
through
DATA ANALYSIS
in
GRADES 6–8

George W. Bright
Wallece Brewer
Kay McClain
Edward S. Mooney

Susan N. Friel
Grades 6–8 Editor

Peggy A. House
Navigations Series Editor

NATIONAL COUNCIL OF
TEACHERS OF MATHEMATICS

Copyright © 2003 by
The National Council of Teachers of Mathematics, Inc.
1906 Association Drive, Reston, VA 20191-1502
(703) 620-9840; (800) 235-7566
www.nctm.org

All rights reserved

ISBN 0-87353-547-2

Printed in the United States of America

TABLE OF CONTENTS

CONTENTS OF CD-ROM

Introduction

Table of Standards and Expectations, Data Analysis and Probability, Grades 6–8

Excel Data Files

Applet Activities

Instructions for Users of Minitools
Minitool 1: Using Case-Value Plots to Compare Data
Minitool 2: Using Dot Plots to Compare Data
Minitool 3: Using Scatterplots to Explore Relationships between Two Variables

Blackline Masters and Templates

Blackline Masters
Centimeter Grid Paper
Half-Centimeter Grid Paper

Readings and Supplemental Materials

Teaching Statistics: What's Average?
Susan N. Friel
The Teaching and Learning of Algorithms in School Mathematics

Understanding Students' Understanding of Graphs
Susan N. Friel, George W. Bright, and Frances R. Curcio
Mathematics Teaching in the Middle School

Sticks to the Roof of Your Mouth?
Susan N. Friel and William T. O'Connor
Mathematics Teaching in the Middle School

Learning Statistics with Technology
Gary Kader and Mike Perry
Mathematics Teaching in the Middle School

Essay

Making and Using Histograms
    George W. Bright, Wallece Brewer, Kay McClain, and Edward S. Mooney

About This Book

Since the publication of *Curriculum and Evaluation Standards for School Mathematics* (National Council of Teachers of Mathematics [NCTM] 1989), increasing attention has been placed on data analysis as a coherent strand of the school mathematics curriculum. *Principles and Standards for School Mathematics* (NCTM 2000) sets forth the Data Analysis and Probability Standard, which presents three clear standards for performance in data analysis (see the margin notes). These three standards overlap the elements of the process of statistical investigation shown in chapter 1.

This book offers activities that further students' understanding of the full range of data analysis as an identifiable strand of mathematics content. The four chapters illustrate the general notion of statistics as a process while also provoking discussions of increasingly complex mathematical issues.

Chapter 1, "Setting the Stage for Data Analysis," extends and deepens students' knowledge of data analysis. The important ideas include the following:

- Data analysis as a process
- Interrogating a data set
- Useful language
- Numerical summaries

This chapter extends what students already know about representing data with graphs and numerical summaries and interpreting the graphs and summaries to make sense of the set of data—that is, to interrogate the data set.

Chapter 2, "Comparing Data Sets with Equal Numbers of Elements," introduces the comparison of data sets when the numbers of data points in the sets are the same (i.e., equal Ns). Such comparisons stretch students' reasoning beyond simple descriptions of a data set. The essential ideas include the following:

- Spread and variability
- Shapes of distributions
- Reasoning about data

One purpose of comparing data sets is to decide if they represent populations that are the same or different. When the data sets to be compared have equal numbers of elements, additive reasoning suffices, but the groundwork is laid in such comparisons for more-sophisticated kinds of reasoning.

Chapter 3, "Comparing Data Sets with Unequal Numbers of Elements," makes the transition to the comparison of data sets with numbers of data points that are not the same (i.e., unequal Ns). Multiplicative reasoning is necessary for interpreting comparisons of data sets with unequal Ns. The important ideas include these:

- Relative frequency
- Box plots
- Multiplicative reasoning

Formulate questions that can be addressed with data and collect, organize, and display relevant data to answer them

Select and use appropriate statistical methods to analyze data

Develop and evaluate inferences and predictions that are based on data

Key to Icons

Principles and Standards

CD-ROM

Blackline Master

Three different icons appear in the book, as shown in the key. One alerts readers to material quoted from *Principles and Standards for School Mathematics,* another points them to supplementary materials on the CD-ROM that accompanies the book, and a third signals the blackline masters and indicates their locations in the appendix.

Chapter 3 helps students develop the techniques necessary for determining whether two data sets represent the same or different populations. Accurate determinations are central to making "inferences and predictions that are based on data" (NCTM 2000, p. 248) rather than on assumptions about the data or the populations from which the data are taken.

Chapter 4, "Exploring Bivariate Data," introduces students to the analysis of data involving two variables. The main ideas include—

- bivariate data, and
- scatterplots and prediction.

Bivariate data are created when two characteristics are measured for each element in a sample. One of the central tasks in understanding bivariate data is inferring whether the characteristics are related.

Each chapter begins with a discussion of the important mathematical ideas that it treats. Then an activity is presented that helps teachers understand what knowledge their students might already have about these ideas. This preassessment activity is followed by other activities, many of which have blackline masters that can be used directly with students. The blackline masters, signaled by an icon, can be found in the appendix, along with solutions to the problems where appropriate. They can also be printed from the CD-ROM that accompanies the book. The CD-ROM, also signaled by an icon, includes resources for professional development and three minitools that provide an alternative way of engaging students with some of the same mathematical ideas presented in the book. Throughout the book, an icon appears in the margin next to references to *Principles and Standards for School Mathematics.*

Teachers can help students learn important ideas about data analysis in many ways. What is most important is to select activities that help students develop a conceptual understanding of the ideas while also making sense of data. Tasks need to be sequenced carefully and coordinated with assessments of students' thinking so that both individual students and the class as a whole can progress to increasingly sophisticated reasoning. Returning to familiar ideas in new contexts is often necessary. For example, when a new representation (e.g., a histogram) is introduced, it may be necessary to return to interrogating a single data set as a means of helping students become comfortable with the ideas related to the new representation. Students must understand each new idea before they can use it in sophisticated ways.

Some of the activities in this book can be conducted without access to technology, but many evolving technologies can assist students in becoming proficient at interrogating data. The accompanying CD-ROM includes Excel files of all the data in the book for the convenience of teachers who want their students to use and manipulate the data on the computer. Students can display the data in a variety of tables and charts, through which they can discover patterns, relationships, recursive and explicit behavior, functions, and limits. Today's statisticians often use computer software to help them conduct "interactive data analysis," an analysis of data that evolves as different representations of the data are displayed and interpreted. Because computers can compute and display data very quickly, the information in a data set can be examined from many different perspectives, each of which might reveal an important aspect of the information in the data. McClain, Cobb, and Gravemeijer (2000) describe some of their efforts at using computer tools to allow students to examine different aspects of a data set quickly. Those tools seemed to influence how the students thought about the data and about the answer to the question posed. The tools appeared to help students be more flexible at using data to answer a question that they saw as reasonable and important.

The use of technology in interrogating data can also reveal students' subtle misunderstandings. For example, making different kinds of graphs is easy with graphing software, so students are often observed making all possible graphs. They often have little concern for whether a particular graph is appropriate for the type of data under study. They might, for instance, generate a histogram for categorical data such as eye color. Teachers can take advantage of such "teachable moments" to explore why some graphs are appropriate and some are inappropriate. In this example, the categorical data do not extend over an entire interval, so it is inappropriate to create a histogram for such data. Learning from such mistakes is an effective way for students to develop good metacognitive skills, such as monitoring their work for accuracy and appropriateness.

NAVIGATIONS SERIES

GRADES 6–8

NAVIGATING *through* DATA ANALYSIS

Introduction

The Data Analysis and Probability Standard in *Principles and Standards for School Mathematics* (NCTM 2000) is an affirmation of a fundamental goal of the mathematics curriculum: to develop critical thinking and sound judgment based on data. These skills are essential not only for a select few but for every informed citizen and consumer. Staggering amounts of information confront us in almost every aspect of contemporary life, and being able to ask good questions, use data wisely, evaluate claims that are based on data, and formulate defensible conclusions in the face of uncertainty have become basic skills in our information age.

In working with data, students encounter and apply ideas that connect directly with those in the other strands of the mathematics curriculum as well as with the mathematical ideas that they regularly meet in other school subjects and in daily life. They can see the relationship between the ideas involved in gathering and interpreting data and those addressed in the other Content Standards—Number and Operations, Algebra, Measurement, and Geometry—as well as in the Process Standards—Reasoning and Proof, Representation, Communication, Connections, and Problem Solving. In the Navigations series, the *Navigating through Data Analysis and Probability* books elaborate the vision of the Data Analysis and Probability Standard outlined in *Principles and Standards*. These books show teachers how to introduce important statistical and probabilistic concepts, how the concepts grow, what to expect students to be able to do and understand during and at the end of each grade band, and how to assess what they know. The books also introduce representative instructional activities that help translate the vision of *Principles and Standards* into classroom practice and student learning.

Fundamental Components of Statistical and Probabilistic Thinking

Principles and Standards sets the Data Analysis and Probability Standard in a developmental context. It envisions teachers as engaging students from a very young age in working directly with data, and it sees this work as continuing, deepening and growing in sophistication and complexity as the students move through school. The expectation is that all students, in an age-appropriate manner, will learn to—

- formulate questions that can be addressed with data and collect, organize, and display relevant data to answer them;
- select and use appropriate statistical methods to analyze data;
- develop and evaluate inferences and predictions that are based on data; and
- understand and apply basic concepts of probability.

Formulating questions that can be addressed with data and collecting, organizing, and displaying relevant data to answer them

No one who has spent any time at all with young children will doubt that they are full of questions. Teachers of young children have many opportunities to nurture their students' innate curiosity while demonstrating to them that they themselves can gather information to answer some of their questions.

At first, children are primarily interested in themselves and their immediate surroundings, and their questions center on such matters as "How many children in our class ride the school bus?" or "What are our favorite flavors of ice cream?" Initially, they may use physical objects to display the answers to their questions, such as a shoe taken from each student and placed appropriately on a graph labeled "The Kinds of Shoes Worn in Kindergarten." Later, they learn other methods of representation using pictures, index cards, sticky notes, or tallies. As children move through the primary grades, their interests expand outward to their surroundings, and their questions become more complex and sophisticated. As that happens, the amount of collectible data grows, and the task of keeping track of the data becomes more challenging. Students then begin to learn the importance of framing good questions and planning carefully how to gather and display their data, and they discover that organizing and ordering data will help uncover many of the answers that they seek. However, learning to refine their questions, planning effective ways to collect data, and deciding on the best ways to organize and display data are skills that children develop only through repeated experiences, frequent discussions, and skillful guidance from their teachers. By good fortune, the primary grades afford many opportunities—often in conjunction with lessons on counting, measurement, numbers, patterns, or other school subjects—for children to pose interesting questions and develop ways of collecting data that will help them formulate answers.

As students move into the upper elementary grades, they will continue to ask questions about themselves and their environment, but their questions will begin to extend to their school or the community or the world beyond. Sometimes, they will collect their own data; at other times, they will use existing data sets from a variety of sources. In either case, they should learn to exercise care in framing their questions and determining what data to collect and when and how to collect them. They should also learn to recognize differences among data-gathering techniques, including observation, measurement, experimentation, and surveying, and they should investigate how the form of the questions that they seek to answer helps determine what data-gathering approaches are appropriate. During these grades, students learn additional ways of representing data. Tables, line plots, bar graphs, and line graphs come into play, and students develop skill in reading, interpreting, and making various representations of data. By examining, comparing, and discussing many examples of data sets and their representations, students will gain important understanding of such matters as the difference between categorical and numerical data, the need to select appropriate scales for the axes of graphs, and the advantages of different data displays for highlighting different aspects of the same data.

During middle school, students move beyond asking and answering the questions about a single population that are common in the earlier years. Instead, they begin posing questions about relationships among several populations or samples or between two variables within a single population. In grades 6–8, students can ask questions that are more complex, such as "Which brand of laundry detergent is the best buy?" or "What effect does light [or water or a particular nutrient] have on the growth of a tomato plant?" They can design experiments that will allow them to collect data to answer their questions, learning in the process the importance of identifying relevant data, controlling variables, and choosing a sample when it is impossible to collect data on every case. In these middle school years, students learn additional ways of representing data, such as with histograms, box plots, or relative-frequency bar graphs, and they investigate how such displays can help them compare sets of data from two or more populations or samples.

By the time students reach high school, they should have had sufficient experience with gathering data to enable them to focus more precisely on such questions of design as whether survey questions are unambiguous, what strategies are optimal for drawing samples, and how randomization can reduce bias in studies. In grades 9–12, students should be expected to design and evaluate surveys, observational studies, and experiments of their own as well as to critique studies reported by others, determining if they are well designed and if the inferences drawn from them are defensible.

Selecting and using appropriate statistical methods to analyze data

Teachers of even very young children should help their students reflect on the displays that they make of the data that they have gathered. Students should always thoughtfully examine their representations to determine what information they convey. Teachers can prompt young children to derive information from data displays through questions like "Do more children in our class prefer vanilla ice cream, or do more prefer chocolate ice cream?" As children try to interpret their work, they come to realize that data must be ordered and organized to convey answers to their questions. They see how information derived from data, such as their ice cream preferences, can be useful—in deciding, for example, how much of particular flavors to buy for a class party. In the primary grades, children ordinarily gather data about whole groups—frequently their own class—but they are mainly interested in individual data entries, such as the marks that represent their own ice cream choices. Nevertheless, as children move through the years from prekindergarten to grade 2, they can be expected to begin questioning the appropriateness of statements that are based on data. For example, they may express doubts about such a statement as "Most second graders take ballet lessons" if they learn that only girls were asked if they go to dancing school. They should also begin to recognize that conclusions drawn about one population may not apply to another. They may discover, for instance, that bubble gum and licorice are popular ice cream flavors among their fellow first graders but suspect that this might not necessarily be the case among their parents.

In contrast with younger children, who focus on individual, often personal, aspects of data sets, students in grades 3–5 can and should be guided to see data sets as wholes, to describe whole sets, and to compare one set with another. Students learn to do this by examining different sets' characteristics—checking, for example, values for which data are concentrated or clustered, values for which there are no data, or values for which data are unusually large or small (*outliers*). Students in these grades should also describe the "shape" of a whole data set, observing how the data spread out to give the set its *range*, and finding that range's center. In grades 3–5, the center of interest is in fact very often a measure of a data set's center—the *median* or, in some cases, the *mode*. In the process of learning to focus on sets of data rather than on individual entries, students should start to develop an understanding of how to select *typical* or *average* (*mean*) values to represent the sets. In examining similarities and differences between two sets, they should explore what the means and the ranges tell about the data. By using standard terms in their discussions, students in grades 3–5 should be building a precise vocabulary for describing the characteristics of the data that they are studying.

By grade 5, students may begin to explore the concept of the mean as a balance point in an informal way, but a formal understanding of the mean and its use in describing data sets does not become important until grades 6–8. By this time, just being able to compute the mean is no longer enough. Students need ample opportunities to develop a fundamental conceptual understanding—for example, by comparing the usefulness and appropriateness of the mean, the median, and the mode as ways of describing data sets in different contexts. In middle school, students should also explore questions that are more probing, such as "What impact does the spread of a distribution have on the value of the mean [or the median]?" Or "What effect does changing one data value [or more than one] have on different measures of center—the mean, the median, and the mode?" Technology, including spreadsheet software, calculators, and graphing software, becomes an important tool in grades 6–8, enabling students to manipulate and control data while they investigate how changes in certain values affect the mean, the median, or the distribution of a set of data. Students in grades 6–8 should also study important characteristics of data sets, such as *symmetry*, *skewness*, and *interquartile range*, and should investigate different types of data displays to discover how a particular representation makes such characteristics more or less apparent.

As these students move on into grades 9–12, they should grow in their ability to construct an appropriate representation for a set of univariate data, describe its shape, and calculate summary statistics. In addition, high school students should study linear transformations of univariate data, investigating, for example, what happens if a constant is added to each data value or if each value is multiplied by a common factor. They should also learn to display and interpret bivariate data and recognize what representations are appropriate under particular conditions. In situations where one variable is categorical—for example, gender—and the other is numerical—a measurement of height, for instance—students might use appropriately paired box plots or histograms to compare the heights of males and females in a given group. By contrast, students who are presented with bivariate numerical data—for example, measurements of height and arm span—might use a scatterplot to represent their data, and they should be able to describe the shape of the scatterplot and use it to analyze the relationship between the two lengths measured—height and arm span. Types of analyses expected of high school students include finding functions that approximate or "fit" a scatterplot, discussing different ways to define "best fit," and comparing several functions to determine which is the best fit for a particular data set. Students should also develop an understanding of new concepts, including *regression*, *regression line*, *correlation*, and *correlation coefficient*. They should be able to explain what each means and should understand clearly that a correlation is not the same as a causal relationship. In grades 9–12, technology that allows users to plot, move, and compare possible regression lines can help students develop a conceptual understanding of residuals and regression lines and can enable them to compute the equation of their selected line of best fit.

Developing and evaluating inferences and predictions that are based on data

Observing, measuring, or surveying every individual in a population is an appropriate way of gathering data to answer selected questions. Such "census data" is all that we expect from very young children, and teachers in the primary grades should be content when their students confine their data gathering and interpretation to their own class or another small group. But as children mature, they begin to understand that a principal reason for gathering and analyzing data is to make inferences and predictions that apply beyond immediately available data sets. To do that requires sampling and other more advanced statistical techniques.

Teachers of young children lay a foundation for later work with inference and prediction when they ask their students whether they think that another group of students would get the same answers from data that they did. After discussing the results of a survey to determine their favorite books, for example, children in one first-grade class might conclude that their peers in the school's other first-grade class would get similar results but that the fourth graders' results might be quite different. The first graders could speculate about why this might be so.

As students move into grades 3–5, they should be expected to expand their ability to draw conclusions, make predictions, and develop arguments based on data. As they gain experience, they should begin to understand how the data that they collect in their own class or school might or might not be representative of a larger population of students. They can begin to compare data from different samples, such as several fifth-grade classes in their own school or other schools in their town or state. They can also begin to explore whether or not samples are representative of the population and identify factors that might affect representativeness. For example, they could consider a question like "Would a survey of children's favorite winter sports get similar results for samples drawn from Colorado, Hawaii, Texas, and Ontario?" Students in the upper grades should also discuss differences in what data from different samples show and factors that might account for the observed results, and they can start developing hypotheses and designing investigations to test their predictions.

It is in the middle grades, however, that students learn to address matters of greater complexity, such as the relationship between two variables in a given population or sample, or the relationships among several populations or samples. Two concepts that are emphasized in grades 6–8 are *linearity* and *proportionality*, both of which are important in developing students' ability to interpret and draw inferences from data. By using scatterplots to represent paired data from a sample—for example, the height and stride length of middle schoolers—students might observe whether the points of the scatterplot approximate a line, and if so, they can attempt to draw the line to fit the data. Using the slope of that line, students can make conjectures about a relationship between height and stride length. Furthermore, they might decide to compare a scatterplot for middle school boys with one for middle school girls to determine if a similar ratio applies for both groups. Or they might draw box plots or relative-frequency histograms to represent data on the heights of samples of middle school boys and high school boys to investigate the variability in height of boys of different ages. With the help of graphing technology, students can examine many data sets and learn to differentiate between linear and nonlinear relationships, as well as to recognize data sets that exhibit no relationship at all. Whenever possible, they should attempt to describe observed relationships mathematically and discuss whether the conjectures that they draw from the sample data might apply to a larger population. From such discussions, students can plan additional investigations to test their conjectures.

As students progress to and through grades 9–12, they can use their growing ability to represent data with regression lines and other mathematical models to make and test predictions. In doing so, they learn that inferences about a population depend on the nature of the samples, and concepts such as *randomness*, *sampling distribution*, and *margin of error* become important. Students will need firsthand experience with many different statistical examples to develop a deep understanding of the powerful ideas of inference and prediction. Often that experience will come through simulations that enable students to perform hands-on experiments while developing a more intuitive understanding of the relationship between characteristics of a sample and the corresponding characteristics of the population from which the sample was drawn. Equipped with the concepts learned through simulations, students can then apply their understanding by analyzing statistical inferences and critiquing reports of data gathered in various contexts, such as product testing, workplace monitoring, or political forecasting.

Understanding and applying basic concepts of probability

Probability is connected to all mathematics from number to geometry. It has an especially close connection to data collection and analysis. Although students are not developmentally ready to study probability in a formal way until much later in the curriculum, they should begin to lay the foundation for that study in the years from prekindergarten to grade 2. For children in these early years, this means informally considering ideas of likelihood and chance, often by thinking about such questions as "Will it be warm tomorrow?" and realizing that the answer may depend on particular conditions, such as where they live or what month it is. Young children also recognize that some things are sure to happen whereas others are impossible, and they begin to develop notions of *more likely* and *less likely* in various everyday contexts. In addition, most children have experience with common devices of chance used in games, such as spinners and dice. Through hands-on experience, they become aware that certain numbers are harder than others to get with two dice and that the pointer on some spinners lands on certain colors more often than on others.

In grades 3–5, students can begin to think about probability as a measurement of the likelihood of an event, and they can translate their earlier ideas of *certain, likely, unlikely,* or *impossible* into quantitative representations using 1, 0, and common fractions. They should also think about events that are neither certain nor impossible, such as getting a 6 on the next roll of a die. They should begin to understand that although they cannot know for certain what will happen in such a case, they can associate with the outcome a fraction that represents the frequency with which they could expect it to occur in many similar situations. They can also use data that they collect to estimate probability—for example, they can use the results of a survey of students' footwear to predict whether the next student to get off the school bus will be wearing brown shoes.

Students in grades 6–8 should have frequent opportunities to relate their growing understanding of proportionality to simple probabilistic situations from which they can develop notions of chance. As they refine their understanding of the chance, or likelihood, that a certain event will occur, they develop a corresponding sense of the likelihood that it will not occur, and from this awareness emerge notions of complementary events, mutually exclusive events, and the relationship between the probability of an event and the probability of its complement. Students should also investigate simple compound events and use tree diagrams, organized lists, or similar descriptive methods to determine probabilities in such situations. Developing students' understanding of important concepts of probability—not merely their ability to compute probabilities—should be the teacher's aim. Ample experience is important, both with hands-on experiments that generate empirical data and with computer simulations that produce large data samples. Students should then apply their understanding of probability and proportionality to make and test conjectures about various chance events, and they should use simulations to help them explore probabilistic situations.

Concepts of probability become increasingly sophisticated during grades 9–12 as students develop an understanding of such important ideas as *sample space, probability distribution, conditional probability, dependent* and *independent events,* and *expected value.* High school students should use simulations to construct probability distributions for sample spaces and apply their results to predict the likelihood of events. They should also learn to compute expected values and apply their knowledge to determine the fairness of a game. Teachers can reasonably expect students at this level to describe and use a sample space to answer questions about conditional probability. The solid understanding of basic ideas of probability that students should be developing in high school requires that teachers show them how probability relates to other topics in mathematics, such as counting techniques, the binomial theorem, and the relationships between functions and the area under their graphs.

Developing a Data Analysis and Probability Curriculum

Principles and Standards reminds us that a curriculum that fosters the development of statistical and probabilistic thinking must be coherent, focused, and well articulated—not merely a collection of lessons or activities devoted to diverse topics in data analysis and probability. Teachers should introduce rudimentary ideas of data and chance deliberately and purposefully in the early years, deepening and expanding their students' understanding of them through frequent experiences and applications as students progress through the curriculum. Students must be continually challenged to learn and apply increasingly sophisticated statistical and probabilistic thinking and to solve problems in a variety of school, home, and real-life settings.

The six *Navigating through Data Analysis and Probability* books make no attempt to present a complete, detailed data analysis and probability curriculum. However, taken together, these books illustrate how selected "big ideas" behind the Data Analysis and Probability Standard develop this strand of the mathematics curriculum from prekindergarten through grade 12. Many of the concepts about data analysis and probability that the books present are closely tied to topics in algebra, geometry, number, and measurement. As a result, the accompanying activities, which have been especially designed to put the Data Analysis and Probability Standard into practice in the classroom, can also reinforce and enhance students' understanding of mathematics in the other strands of the curriculum, and vice versa.

Because the methods and ideas of data analysis and probability are indispensable components of mathematical literacy in contemporary life, this strand of the curriculum is central to the vision of mathematics education set forth in *Principles and Standards for School Mathematics*. Accordingly, the *Navigating through Data Analysis and Probability* books are offered to educators as guides for setting successful courses for the implementation of this important Standard.

NAVIGATING *through* DATA ANALYSIS

Chapter 1
Setting the Stage for Data Analysis

Important Mathematical Ideas

Much of the current work in helping students understand statistical reasoning involves conceptualizing data analysis as a process involving collecting data, describing and presenting data using statistical methods, and drawing conclusions from data (Moore 1991). *Principles and Standards for School Mathematics* (National Council of Teachers of Mathematics [NCTM] 2000, p. 248) sets out three general standards related to the performance of middle-grades students in data analysis:

- Formulate questions that can be addressed with data and collect, organize, and display relevant data to answer them
- Select and use appropriate statistical methods to analyze data
- Develop and evaluate inferences and predictions that are based on data

Middle-grades students are expected to understand the four major components of statistics as a process: pose questions, gather data, analyze data, and interpret data. The interaction of these components can be diagrammed in several different ways; figure 1.1 displays a variation by Kader and Perry (1997), which builds on Graham's (1987) model. This diagram illustrates two major points. First, the problem and the data set are crucial elements in a statistical investigation. The interpretation of the results is filtered through the problem context as part of the process of sense making. Second, interpreting the results may lead to a reexamination of the questions posed. Frequently, more questions will come to

One example of engaging students in a motivating problem is given by VanLeuvan (1997).

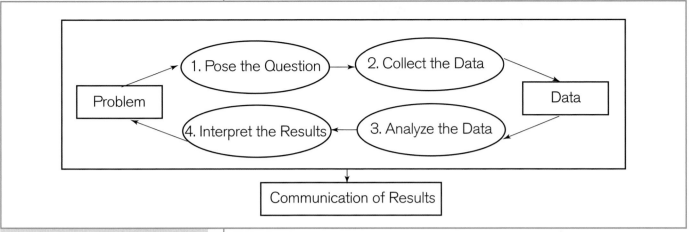

mind, but at the very least, the relationship between the questions posed and the data gathered will be understood more deeply as a result of interpreting the results.

Students are motivated by decision making, which requires that they recognize and interconnect the various aspects of statistics as a process. Good decisions (i.e., defensible decisions) are made only after students understand the problem, how the data have been gathered, the reasons for the use of a particular means of data collection, how various representations of the data aid in understanding the data, and what interpretations might legitimately be made of the data in relation to the problem.

The activities that follow exemplify how teachers can help students develop an understanding of statistics as a process. These tasks should be accompanied by a careful assessment and a thorough discussion of both students' solutions and the thinking that led to their solutions. Through such discourse, students can reflect on their thinking, partly by making their thinking public and partly by comparing their own understandings with the understandings of others. Developing meaning through discussions among students and their teacher lays the foundation for the development of rich conceptual networks that connect statistical ideas to one another and to other ideas in mathematics, such as proportional reasoning.

Interrogating a data set

One of the images that has influenced the structure of this book is that of the student "interrogating a data set"—that is, making sense of a set of data in a particular context. The investigator wants the data to "tell" him or her everything that they can. Of course, a data set is static and takes no initiative in revealing its secrets. The person trying to understand a data set makes decisions about how the data can best be organized and represented. The question to be answered must then be reexamined in light of the information revealed by the organization or representation of the data. This reexamination, in turn, may lead the investigator to consider whether the available data are adequate for answering the question and whether any preliminary analysis points to an acceptable answer to the question. The notion of interrogating the data, then, involves the entire process of a statistical investigation.

McClain (1999) provides some examples of students interrogating a data set.

In interrogating a data set, students are very likely to make some kind of graph of the data. It is important that they understand that the graphs used for discrete data are different from those used for continuous data. Discrete data are typically represented by counts, either of numerical values (e.g., number of siblings) or of nonnumerical values (e.g., eye color); these data can be represented with bar graphs or (possibly) circle graphs. Continuous data are typically measurements (e.g., heights of students); these data can be represented with almost any kind of graph. (The chart in fig. 1.2 summarizes essential information about the kinds of graphs that middle-grades students might use; it includes an example of each type.) Discussing students' representations of data of both types can help the students develop a sense of which graphs are appropriate for different kinds of data.

You may notice mistakes that students make in constructing various types of graphs. It is important that the students realize when their representations are not correct, but it is unlikely that simply telling them what is wrong or how to fix the graph will be effective in the long term. A discussion about a variety of displays and about how those displays help reveal information relevant to the question seems much more likely to generate an attitude of self-assessment.

When students gather their own data, they are likely to have a deeper understanding of the context than when data are simply "handed" to them. However, in certain instances, many of which involve problems that are appealing to middle-grades students, it is impossible for students to gather data. In such cases, an initial discussion of how the data might have been gathered is an important first step in engaging students in interrogating the data. For example, in order to understand and interpret graphs and tables in the popular press (e.g., *USA Today*), students would benefit from a discussion of how the data might have been gathered as well as of what biases might be inherent in the data-collection process.

Useful language

As students enter the middle grades, their understanding of statistics may be limited to making graphs, reading specific bits of information from graphs, and computing measures of center (i.e., mean, median, and mode). As they build on these ideas, it is important that they begin to develop a language—both written and oral—for expressing their growing understanding of statistics. Asking students to describe a set of data by relating it to a specific context is a first step in developing that language. Many students will think that making a graph, computing a mean, and predicting a trend in a data set are relatively independent activities. They need to see the interrelationships among these and other activities that can be carried out for a particular data set. Asking students simply to "describe" a data set is a worthwhile exercise, provided that the students are encouraged to make connections among the various patterns that they observe.

Some specific terminology, such as *cluster of data*, not only may have some intuitive meaning from common usage but also can be related specifically to important mathematical attributes of a data set. It is important to use terminology in a variety of contexts so that students can generalize appropriate uses of the words. Suppose, for example,

A *categorical variable* has values that are labels for a particular attribute (e.g., eye color). A *quantitative variable* has values that not only are numerical but also allow descriptions such as mean and range to be meaningful (e.g., the heights of seventh-grade boys). A *discrete variable* has only countable values (e.g., the number of pets in a home). A *continuous variable* has numerical values that can be any of the values in a range of numbers (e.g., the speed of a car).

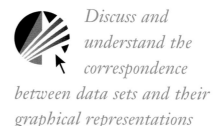

Discuss and understand the correspondence between data sets and their graphical representations

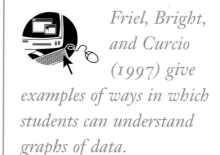

Friel, Bright, and Curcio (1997) give examples of ways in which students can understand graphs of data.

Fig. **1.2.**

Types of graphs

Different graphs are used in different situations; each has both advantages and disadvantages. For example, some graphs are useful for small data sets, whereas others are useful for large data sets. Some graphs display each data value individually, but others "hide" individual values in bars or other visual elements. This chart contains important information about the graphs that middle-grades students are likely to use.

Type of Graph

A *line plot* is a fast way to organize data. The possible data values are listed on a horizontal axis, and one *X* for each element in the data set is placed above the corresponding value. This display works best when the data set has fewer than twenty-five elements and when the range of possible values is not too great. A *dot plot* is similar to a line plot; small dots are used instead of *X*s.

(Landwehr 1986, p. 5)

A *bar graph* shows the frequencies of specific data values in a data set. It can be used for categorical or numerical data, but it is one of the most common ways to display categorical data. The length of the bar drawn for each data value represents the frequency of that value. Bars may be drawn vertically or horizontally. To avoid confusion, the bars should be the same width. In elementary school mathematics, a *case-value plot* is sometimes created. In a case-value plot, the height of the bar drawn for each data element represents the data value. Bar graphs and case-value plots are not interpreted in the same ways, and sometimes students confuse the interpretation of these two displays.

(Moore 1991, pp. 184–85)

Example

Line Plot

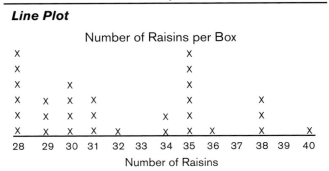

Dot Plot

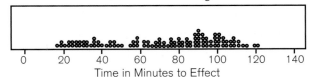

Bar Graph

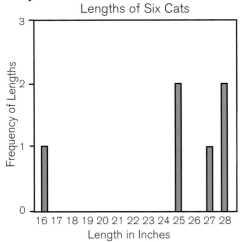

Case-Value Plot

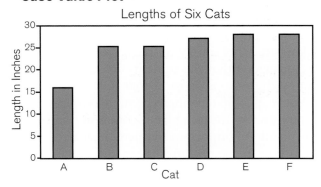

Fig. **1.2.**

Types of graphs

Type of Graph	Example

A *circle graph,* or *pie chart,* is a circle divided into parts, or sectors or wedges. Each part shows the percent of the data elements that are categorized similarly (e.g., grouped into intervals). The parts must sum to 100 percent. Circle graphs are often difficult to make, since each percent must be converted to an angle (i.e., the appropriate fraction of 360°) and the angles are sometimes difficult to draw.

(Moore 1991, pp. 180–81)

Circle Graph

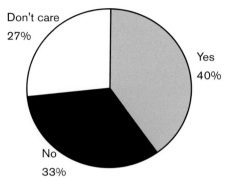

Responses to Julio's Question

A *stem plot* (also called a *stem-and-leaf plot*) is a display that is most often used to "separate" the tens digits from the ones digits of the data values. The tens digits are called the *stems,* and the ones digits are called the *leaves.* Each leaf represents one of the data elements. Ordering the leaves on each stem from least to greatest often facilitates the interpretation of this display. This display works best when the data set contains more than twenty-five elements and when the data values span several decades of values. A stem plot can also be adapted to show simple decimal values—for example, whole numbers and tenths. A back-to-back stem plot can be used to compare two data sets.

(Landwehr 1986, pp. 7–9, 33)

Stem Plot

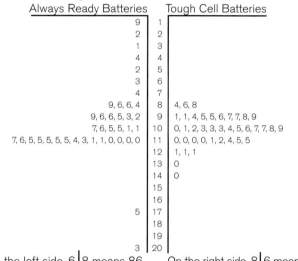

Battery Life (in Hours) for Two Brands

A *histogram* is used when data elements could assume any value in a range—heights or weights of people, for example. The data are organized in equal intervals; the data values are marked on the horizontal axis. Bars of equal width are drawn for each interval, with the height of each bar representing either the number of elements or the percent of elements in that interval; the number or percent is marked on the vertical axis. The bars are drawn without any space between them.

(Moore 1991, pp. 191–92)

Histogram

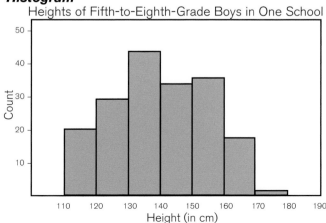

Heights of Fifth-to-Eighth-Grade Boys in One School

Continued

Fig. **1.2.**

Types of graphs

Type of Graph	Example

A *box plot* (also called a *box-and-whiskers plot*) is constructed by marking the "five-point summary" (i.e., the least and greatest values, the median, and the first and third quartiles), drawing a box to capture the interval from the first to the third quartile, and connecting the box to the least and greatest values with line segments. The data elements are not displayed individually, which makes it impossible to determine if there are gaps or clusters in the data. Box plots are very useful, however, for comparing data sets, especially when the data sets are large or when they have different numbers of data elements.

(Landwehr 1986, pp. 57, 73)

Box Plot

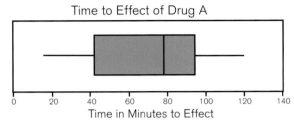

A *line graph* is typically used for continuous data to show the change in a variable—over time, for example. The time is marked on the horizontal axis, and the values of the variable are marked on the vertical axis. Each element of the sample is associated with a value for time and a value of the variable. Each pair of values is graphed, and the points are connected with line segments. It is important to look carefully at the scale marked on the vertical axis, since changing the scale of the vertical axis can dramatically change the visual impression of the graph.

(Moore 1991, pp. 181–83)

Line Graph

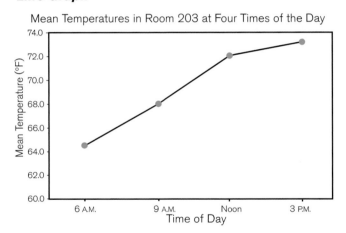

A *scatterplot* is used when two measurements are made for each element of the sample. The graph consists of points on a two-dimensional grid; the two coordinates of each point are determined by the two measurements for the corresponding element of the sample. A scatterplot is one of the best ways to determine if two characteristics are related.

(Landwehr 1986, pp. 84–86, 137)

Scatterplot

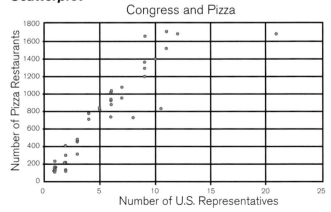

that students gathered data on the amount of time they watched TV (see fig. 1.3). They might say that the graph looks like a hill (or a valley). They might notice that the cluster from five hours to eight hours represents an average of about one hour of TV watching per day, whereas the cluster from thirteen hours to fifteen hours represents an average of about two hours per day. They might focus attention on the gap in the data between twenty-two hours and thirty hours. You can discuss different combinations of hours each day that might result in averages of one to two hours per day (e.g., seven hours on Saturday and no other TV during the week, two hours each on Monday through Thursday and no TV on other days).

Central to developing students' statistical terminology is discourse—both written and oral. By sharing solutions orally in class, students will begin to see the need for precision in language. An explanation that seems crystal clear to the person speaking will often seem confusing to other students. The use of clarifying and probing questions will help the speaker reformulate the explanation in ways that other students in the class can understand. Sometimes after a clear explanation has been given, it is useful to ask the speaker to repeat it so that it gets stored in memory. It may then be easier for that person to write the explanation. The process of clarifying oral language and then moving to written language may help students write their mathematical ideas in ways that communicate the ideas clearly to readers.

You might also discuss "what if" questions. What would happen to the data display if everyone watched one more hour of TV per week? Would the shape change? Would the clusters change? (No, but the entire distribution would be shifted one unit to the right.) What would happen if everyone watched twice as much TV? Would the shape change? Would the clusters change? (The distribution would be "stretched out horizontally," so the clusters might appear somewhat less dense.)

Interrogating data involves comprehending the data set as a whole. Questioning—either of the students by the teacher or of the data by the students—is an important way to develop comprehension. There is considerable agreement about three levels of questions that graphs may be used to answer: "an elementary level focused on extracting data from a graph (i.e., locating, translating); an intermediate level characterized by interpolating and finding relationships in the data as shown on a graph (i.e., integrating, interpreting), and an advanced level that

Fig. **1.3.**

A line plot of the amount of TV watching by middle-grades students for one week

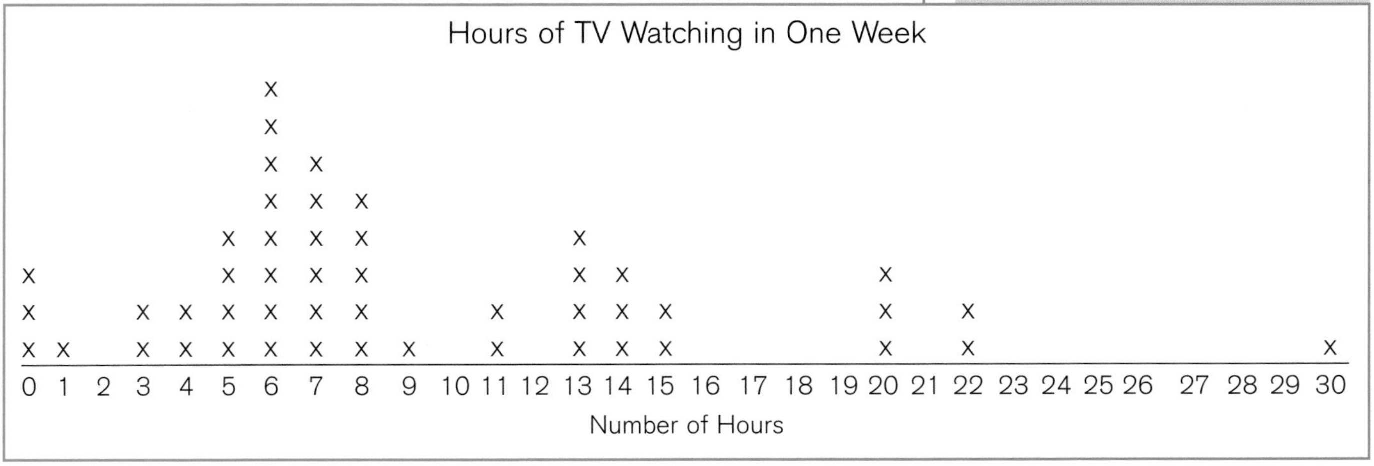

Hours of TV Watching in One Week

requires extrapolating from the data and analyzing the relationships implicit in a graph (i.e., generating, predicting). At the third level, questions provoke students' understanding of the deep structure of the data presented" (Friel, Curcio, and Bright 2001, p. 130). Curcio's (1987) terminology— *read the data*, *read between the data*, and *read beyond the data*—captures the essence of these distinctions. Clearly, as the sophistication of a question increases, the difficulty of answering the question also increases.

Numerical summaries

Numerical summaries—for example, mean, median, or mode—can offer a single value that is representative of a data set. Each of these values reveals something different about the data. Numerical summaries are most effectively used along with other descriptive language. Deciding which of these values to use depends on the nature of the data.

The *mean* of a set of data is the arithmetic average; it is calculated by dividing the sum of the elements by the number of elements. The *median* of a set of data is the middle element after the elements have been ordered by magnitude. For an even number of data, the median is the mean of the middle two elements. The *mode* of a set of data is the most frequent data value. The mode can be used to describe either categorical or quantitative data. The mean and median can be used only for numerical data. The *range* of a data set is the difference between the greatest and the least values in the set. Students are likely to know how to compute the range, although they may not understand its significance. Relating the range to numerical summaries can lay the groundwork for deeper understanding.

When asked to identify a typical value, students often choose the mode. In a standard bar graph, the mode is the value associated with the tallest bar, so it "stands out" in the display. Unless the mode is considerably greater than other values, however, it is often not a particularly representative value.

The mean, or arithmetic average, is the value that is often used when sophisticated mathematical analyses are required. However, it is not intuitively obvious to students that the mean is a "good" value to represent a data set. The mean is highly influenced by extreme values, whereas the median is not. If the data display a few extreme values, the mean may not be as representative as the median. The mean, however, has some important mathematical properties that make it very useful for more-advanced statistical analyses.

The median is, in some sense, the least visible of these values; in order to identify it, the data have to be ordered, which students are unlikely to do without some prodding. The median is not influenced by one or two extreme values, so when the data display outliers, the median may be a better representative of the data set than the mean. Students need to gain experience with both the mean and the median.

Discussions about which numerical summary is best for a particular data set can reveal important information about how students are thinking about data. Do they have a sense of the entire data set, or is their focus only on particular values?

Friel (1998) suggests ways to help students develop an understanding of the mean.

Navigating through Data Analysis and Probability in Grades 3–5 (Chapin et al. 2002) offers ideas for introducing measures of center.

Find, use, and interpret measures of center and spread, including mean

What Might Students Already Know about These Ideas?

The activity Lengths of Cats assesses students' understanding of data presented in a graph. This activity does not address all aspects of statistics as a process. Rather, it reveals whether students can read data from a graph, manipulate the data, and identify trends in the data. Knowing how well students succeed in this activity can be very useful in planning future instruction.

Lengths of Cats

Goal

To assess students'—

- ability to interpret information presented in a bar graph;
- ability to explain the reasoning behind their answers;
- skill at reading data, reading between data, and reading beyond data in a graph.

Materials

- A copy of the blackline master "Lengths of Cats" for each student

p. 84

Activity

Set up the activity by discussing what kinds of pets and how many pets the students have. Cats are popular pets, so it is reasonable to assume that students might be interested in their fellow students' cats. The students might want to know what the typical pet cat looks like.

Discuss what characteristics of cats the class might examine. The students are likely to suggest weight, breed, color, and age; help them expand the list by suggesting such attributes as length, eye color, and pad color. Then discuss how they might gather data—by describing their own cats, surveying all the cats in their neighborhood, searching for information on the Internet, and contacting veterinarians. The data used in this activity are from a survey of students. The variable chosen for study was the length of the cat from the nose to the tip of the tail. The students measured their cats to generate the data in this activity.

Some questions in this activity are designed to probe students' understanding of the bar graph itself; others, to probe students' understanding of the data set and to determine if students can respond on the basis of the information displayed in a graph of the data or if they rely on personal experiences with cats and ignore the data. An analysis of the students' answers may reveal how well they understand the conventions of bar graphs and what information they consider relevant in the graph or in the data set. This information is useful for planning future instruction in statistics and statistical reasoning.

Distribute a copy of "Lengths of Cats" to each student. The students should work individually, although you may want them to compare answers with a partner before you discuss the activity with the whole class. This activity requires students to read a graph and complete tasks requiring different skills: reading the data (question 1), reading between the data (questions 2 and 3), and reading beyond the data (questions 4 and 5). During the discussion, ask the students to think about the differences in the kinds of questions that were asked. The students should recognize that some questions ask simply for information from the graph, whereas other questions ask for interpretations of the data displayed in the graph.

Discussion

The first question asks the students to read data directly from the graph—namely, that the bar above 30 is three units high. That is, three cats measured thirty inches long. Some students may say that there is only one bar for thirty inches, so only one cat is thirty inches long. This response may indicate confusion about what a bar represents.

The second question requires students to add all the frequencies represented in the bars; there are twenty-five cats in all. One common error students make is to count only the number of bars (i.e., 12).

The third question often yields a variety of answers. The correct answer (66) is the sum of the lengths—16, 25, and 25; since the shortest cat is sixteen inches long and the next two cats are twenty-five inches long, the lengths of the three shortest cats are sixteen, twenty-five, and twenty-five inches. A common error is to add 16, 25, and 27, the lengths associated with the three leftmost bars. Another common error is to add the numbers associated with the first three shortest bars, namely, 16, 27, and 32. Some students who answer question 1 correctly do not answer question 3 correctly, which suggests a need to help them connect reading data from a graph (almost a rote skill) with relating the data to the problem. One technique for doing so is to make a frequency table for the data.

Questions 4 and 5 ask students to find patterns in the data taken as a whole, and question 5 asks them to make predictions from those patterns. When asked about what is typical, students often focus on the mode (here, 31 and 33). It is probably more reasonable to focus on the cluster of data between 28 and 31 (or between 27 and 33) and indicate that a typical cat would fall somewhere in that interval. However, some students may give the answer "2 inches" because there are five bars of height 2, and they confuse the representation of frequency with the representation of the length of the cats. (The most frequent height of the bars is two units in this example.) This error sometimes derives from a confusion between a frequency bar graph (like the one on the activity sheet) and a case-value plot, typical in the primary grades, in which each data element has its own bar.

Several different answers that students might give to question 5 are acceptable. The question is intentionally "open" in order to reveal

A *case* is an individual thing (e.g., person, animal) for which values of variables are recorded. A *case-value plot* (e.g., the graph in fig. 1.4) shows the values of individual cases; cat length is shown on the vertical axis. On a standard bar graph (e.g., the graph in fig. 1.5), cat length is shown on the horizontal axis and frequency is indicated on the vertical axis.

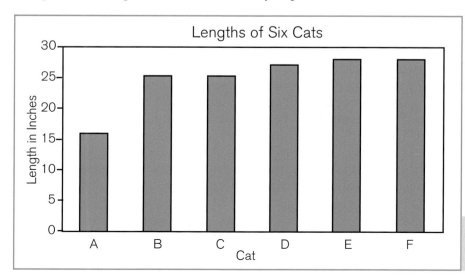

Fig. **1.4.**

A case-value plot of the lengths of six cats

Fig. **1.5.**

A bar graph of the lengths of six cats

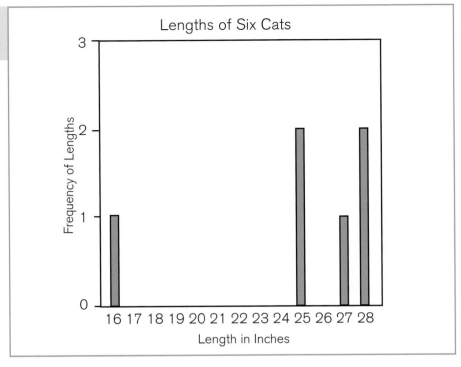

students' conceptual understanding and how they reason about data. The ideal answers would build on the patterns identified in question 4, but often students' responses do not reflect their previous answer. An acceptable answer would be that the typical cat is between twenty-eight and thirty-one inches long, although a somewhat larger interval would also be acceptable. A larger interval would "capture" a greater percent of the data. Some students may erroneously conclude that the next cat will be thirty-four inches ("because there aren't any cats there yet") or thirty-eight inches ("because that is the next number on the *x*-axis").

This activity might be used to determine the degree of consistency that each student displays in the answers to the five questions. That is, is the understanding revealed by the answers to the first two or three questions used to build answers to the last two questions?

Selected Instructional Activities

Activities should give students opportunities to be successful while supporting them in developing an increasingly sophisticated understanding of interrogating a data set. Activities can focus on different aspects of the data-analysis process, such as making a graph (e.g., the activity TV Watching), relating numerical summaries to data (e.g., Making the Data), or interpreting graphs (e.g., Drop Off). With these kinds of exercises as a basis, teachers can help students develop their understanding and skills in many different ways.

TV Watching

Goals

- Make an appropriate graph for discrete data
- Describe the characteristics (e.g., clustering of data) of the graph
- Use the shape of the graph to draw conclusions about the data

Materials

- A copy of the blackline master "TV Watching" for each student or pair of students
- Half-centimeter grid paper (available on the CD-ROM)

p. 85

Activity

Introduce the activity by asking, "Do you think middle-grades students watch too much TV?" Some students may claim that they don't watch too much but that other students do. Some students may ask what is meant by *too much*. Then ask, "What kind of information might help us answer the question?" Help the students see that the first step is to find out how much TV students actually watch. Suggest that a survey might be one way of getting that information.

Distribute the grid paper and a copy of "TV Watching" to each student or pair of students, and introduce the tasks by pointing out that the data on the activity sheet are from two classes of students. Call attention to a couple of the values and ask someone to describe what the values represent. You may need to clarify that for these data, a week consists of seven days. The value 7, then, might be distributed differently throughout a week; for example, the seven hours might represent (*a*) exactly one hour of TV each day of the week, (*b*) three hours on Saturday, four hours on Sunday, and no TV on other days, (*c*) seven hours on Thursday and no TV on other days, or (*d*) some other combination. Each of these possibilities would result in an average of one hour of TV each day. The value 14 might indicate that the student watched TV seven hours on Sunday and again on Saturday or any other combination that totals fourteen hours per week. Each of these possibilities would result in an average of two hours of TV each day. The value 0, however, means that the student did not watch any TV at all during the week; there is no other possibility. You might also ask the students to think about the number of hours they watched TV last week and then decide if their values are in the data set. As an alternative or an extension to this task, you can ask students to gather their own data about TV-watching habits.

Discuss briefly why none of the students reported any time other than a whole number of hours. (Perhaps the students agreed to round their times to the nearest hour.) Tell the students that they will make a graph of the data that could help the parents of the students surveyed decide if their children are watching too much TV. Ask, "Might the graph that you are going to draw be different if the students had reported hours and minutes?" (Yes, it almost certainly would be different, although it is impossible to know exactly *how* the graphs would differ.)

The students may work on the tasks either individually or with partners. The first task requires the students to represent the data in a graph as a means of interrogating the data easily. The second task helps the students translate the hours per week into the approximate number of hours per day. Doing so may help the students decide whether the amounts of TV watching are acceptable. The third task leads the students to view the data from the perspective of the parents of the students. Changing the perspective should help the students examine the data more deeply.

Discussion

The students might construct a variety of graphs, but they are most likely to use either a line plot, a bar graph, or a stem plot. (See Curcio [2001] for a discussion of how students can make those types of graphs.) A bar graph for these data is shown in figure 1.6; a line plot is shown in figure 1.3. The biggest cluster of data is from 5 to 8, although some students will focus on the wider interval of 3 through 9. A second, smaller cluster occurs from 13 to 15.

Fig. 1.6.

A bar graph of the amount of TV watching by middle-grades students for one week

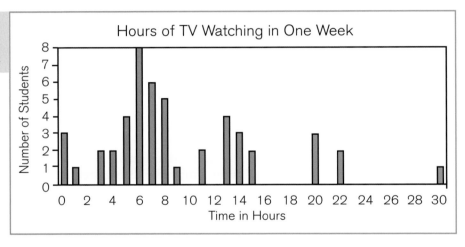

Some students may compute the mean (approximately 9.3 hours), median (7 hours), or mode (6 hours). Since the median and mode are close in value, the students may decide that six or seven hours a week is a representative value for these data. The value of the mean falls in "the hole" between nine and eleven hours, so it does not match any of the data values. Therefore the students may erroneously determine that the mean is not representative and decide to ignore it.

If you ask, "What do you notice about the number of hours these students watch TV?" the class may notice that about half the students watch about seven hours of television a week, or an average of about one hour a day. About a fifth of the students watch about fourteen hours a week, or an average of about two hours a day. A few students watch no TV, and a few watch more than twenty-one hours a week, or an average of more than three hours a day. The students might conclude that the parents would be comfortable with one hour per day, but there may be some disagreement about whether the parents would approve of two or three hours per day. Such disagreements can serve as a springboard for a discussion of how watching too much TV can interfere with other kinds of activities (e.g., studying). The thing to look for, however, is how the students use the shape of the data (e.g., clusters) to justify their conclusions about what parents might say. For example, for the bar graph, students might focus on the groupings of the taller bars. If they observe, instead, "My parents would say ...," ask them to think about the parents of the students represented by the data rather than their own parents: "What do you think the parents of these students might say?"

A final discussion might focus on whether the data were collected in a reliable or representative way. You might focus in particular on events that might have made the previous week's TV watching representative or unrepresentative of a longer period, such as an entire school year. The students might raise concerns such as the following:

- Was the previous week a school week or a vacation week?
- Is the students' recall of TV watching accurate?
- Was the TV watching constant across the days, or was it more concentrated on certain days?

The students may have used measures of center—mean, median, or mode—to describe the TV-watching data. They may not, however, understand how these values are affected by the data values. The next activity, Making the Data, will help them connect data distributions with particular values of these measures of center.

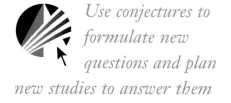

Use conjectures to formulate new questions and plan new studies to answer them

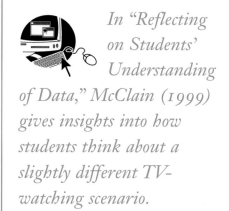

In "Reflecting on Students' Understanding of Data," McClain (1999) gives insights into how students think about a slightly different TV-watching scenario.

Making the Data

Goals

- Construct a data set using summary information about the data
- Identify differences in data sets that share some common summary information
- Explain and use different strategies for constructing a data set

p. 86

Materials

- A copy of the blackline master "Making the Data" for each student

Activity

This activity will help students deepen their understanding of mean, median, and range by exploring some relationships among these numerical summaries. Distribute a copy of "Making the Data" to each student. The students should work individually on the first three questions. Then you might ask them to compare their answers with a partner's. In many cases, the answers will be different; if so, the students should explain to each other why they think their answers are correct. TV watching is again the context in this activity, so students who have completed the previous activity will find it familiar.

The activity requires students to create data sets as more restrictions are placed on the values. First, only the mean is specified; then both the mean and the median are specified; and finally the mean, median, and range are specified. Even the first task may be difficult for some students. To help these students, you can suggest that they draw eight blank underscores for the eight numbers that they need to supply.

Before the students work on task 4, conduct a whole-class discussion of some of the answers to tasks 1–3 and the strategies that the students used to arrive at those answers. You may have at least one student who recognizes that the product of the number of data elements and the mean is the sum of the data elements. If not, you might say, "A student in my first-period class said that the sum of the elements should be the same as the product of the mean and the number of data elements. Is she correct?" Be sure that a strategy based on this information is shared during the discussion, but do not start with it. First, ask some students who used trial and error to explain how they found their data and how they adjusted the data as more information about the numerical summaries was added in tasks 2 and 3.

Discussion

Textbooks often require students to compute various statistics (e.g., mean, median) for a given set of numbers, but it is much less common to ask students to make up data that would have resulted in particular statistical values. The most common strategy that the students are likely to use for such problems is trial and error. However, this strategy is often not the most efficient, so it is important to ask different students to share their strategies for solving the problems. A few students may use strategies that are new to you.

Find, use, and interpret measures of center and spread, including mean

For task 1, since there are eight numbers and their mean is 5, the total of all the values must be 40. Indeed, any set of eight numbers whose sum is 40 will have a mean of 5. Of course, explaining this strategy to the students reduces this activity to a rote exercise, but it is important to help them see the relationship between the number of data elements (8), the sum of the elements (40), and the mean (5). Even students who can compute means quickly and accurately often do not see this relationship immediately. A popular strategy is to "balance" the numbers around the mean. For this problem, the students might start with 4 and 6 (4 = 5 − 1 and 6 = 5 + 1), then write 3 and 7 (3 = 5 − 2 and 7 = 5 + 2), then write 2 and 8, and finally write 1 and 9. Data generated in this way will be symmetric about the mean; a discussion of this strategy should help students understand this point explicitly. Each of the pairs of numbers has a mean of 5, and students often intuit that the mean of all four pairs must also be 5. You should challenge the students to explain why that is so—for instance, by asking, "If I have two data sets with the same mean and I put them together, will the mean of the combined data set be the same as the means of the individual data sets? Why?" The mean of the combined data set will be the same because the sum of the elements in each data set is the mean times the number of data elements:

In the activity Exploring the Mean in Navigating through Data Analysis and Probability in Grades 3–5 *(Chapin et al. 2002), students explore the mean as a balance point and create data sets for specific means.*

$$\text{Sum of elements} = (\text{Mean}) \times (\text{Number of elements})$$

The mean for the combined data set is

$$\frac{(\text{Sum of elements for set 1}) + (\text{Sum of elements for set 2})}{(\text{Number of elements for set 1}) + (\text{Number of elements for set 2})}$$

Substituting [(Mean) × (Number of elements)] for (Sum of elements) for each data set yields the mean for the combined data set:

Mean of combined data set

$$= \frac{(\text{Mean}) \times (\text{Number of elements for set 1}) + (\text{Mean}) \times (\text{Number of elements for set 2})}{(\text{Number of elements for set 1}) + (\text{Number of elements for set 2})}$$

$$= \frac{(\text{Mean}) \times [(\text{Number of elements for set 1}) + (\text{Number of elements for set 2})]}{(\text{Number of elements for set 1}) + (\text{Number of elements for set 2})}$$

The last expression is the common mean, since the other values can be canceled. This argument is fairly sophisticated, and the form of the explanation above requires some knowledge of symbolic manipulation, so some students may not follow the argument easily. It is interesting, however, to stretch students' thinking about this rather intuitive idea.

One "simple" data set for which the mean is 5 is a set of eight 5s. A few students may ask if they may repeat values. Of course, they may, since the problem does not mention any restrictions on the data values. If the students do use eight 5s, ask them to write a second set for which the values are not all the same. Another simple data set is seven 0s and one 40 (i.e., 0, 0, 0, 0, 0, 0, 0, 40). The data values could be fractions or decimals, although it is unlikely that the students will suggest these possibilities without being prompted.

For question 2, the median, or "middle number," would ideally be 4, but in a set of eight numbers, no single middle number is possible. Therefore, the average of the middle two numbers must be 4. Possibilities for the middle numbers are two 4s, a 3 and a 5, or two other numbers that sum to 8. Students who used a balancing strategy for question 1 may have trouble adjusting the data so that the median is 4. Since the conditions of the problem require that the mean and median be different, the data will not be symmetric about the mean.

For question 3, some students may have great difficulty coordinating all three requirements. As they try to adjust the data to make the range 7, they are likely to end up with a wrong value for the mean or the median. In that case, they need to be encouraged to do more experimenting to adjust the three values. It is quite likely that almost all students will need to use a trial-and-error strategy to be sure that the data values they identify meet all the conditions.

Some possible responses to the first three tasks are given in the solutions section of the appendix. Many other answers would also be acceptable.

Once the students have made connections among the ideas of measures of center, it is important that they relate that knowledge to graphical representations of data. The following activity, Drop Off, asks students to identify which of three given values is the mean, which is the median, and which is the mode. Students should be able to relate the values to the shape of the graph of the data.

Drop Off

Goal

- Interpret information presented in a histogram
- Identify which of given values is the mean, the median, and the mode
- Make inferences from a graph of data

Materials

- A copy of the blackline master "Drop Off" for each student

p. 87

Activity

Many middle-grades students are interested in roller coasters—often, the scarier the better. This activity deals with the maximum drop of fifty-five roller coasters in the United States. The data are presented in the histogram in figure 1.7. For this activity, the students must be able to read the information in a histogram. Thus, you may need to discuss what the width and the height of a bar in a histogram represent. For example, the leftmost bar shows that one roller coaster has a maximum drop of sixty to seventy feet. It is impossible to tell from the graph what the exact maximum drop is for that roller coaster. The next bar to the right shows that six roller coasters have a maximum drop of seventy to eighty feet.

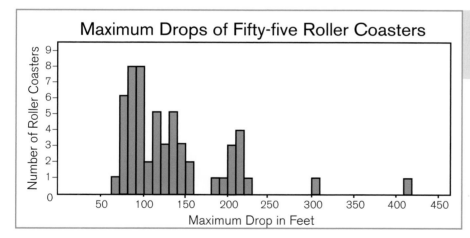

Maximum Drops of Fifty-five Roller Coasters

Fig. **1.7.**

A histogram of the maximum drops of fifty-five roller coasters in the United States

Distribute a copy of "Drop Off" to each student. The students' analyses may be better if they work with partners or in small groups than if they work alone. Group work would be especially helpful for question 5.

Discussion

The first three questions involve students' reading and understanding the data shown in the graph. Twenty-three roller coasters have a maximum drop of less than 100 feet; the sum of the frequencies of the four leftmost bars is $1 + 6 + 8 + 8 = 23$. Some students may say the answer is four, since there are four bars to the left of 100; they may be confusing a bar graph with a case-value plot. Other students may say the answer is twenty-five; they may be including the fifth bar, since its left end point is 100.

In answering question 2, some students may say simply that there are no data between 160 and 180, but they might not be able to draw a conclusion about what this lack of data tells them about the roller coasters—namely, that none of the roller coasters has a maximum drop in this interval. The gap may not be related to anything significant about the design of roller coasters, but if the dates of construction of the coasters were available and if they indicated that the newer coasters had longer maximum drops, the data might indicate that designers deliberately increased the maximum drops in order to make the newer coasters scarier. Question 3 calls attention to the outliers in the data. Of course, the roller coasters represented by the outliers would be the scariest of all.

Question 4 requires students to determine the difference between a "typical" roller coaster in the large data cluster and a typical roller coaster in the small cluster. A roller coaster in the small cluster has a greater maximum drop than one in the large cluster, so it would be scarier for most riders.

To answer question 5, the students should use information about the shape of the data. The median is the middle of the fifty-five values, so the students can count the values to determine that the bar over the interval 110–119 contains the twenty-eighth data value. By a process of elimination, they then know that the median must be 115 instead of 132, the other choice in question 5. The right part of the distribution has two extreme values, which skew the mean toward these values; therefore, the mean must be greater than the median, so it must be 132.

Conclusion

The activities in this chapter are intended to help students integrate and extend what they know about data analysis. One of the important ideas presented is interrogating a data set. If you can help students learn how to interrogate a data set, they will find the process of comparing data sets—with either equal or different numbers of data values—much easier. Helping students increase their flexibility in thinking about what graphs and summary statistics reveal about a data set is a good first step. Gradually increasing the complexity of the problems that you present to students also increases the sophistication of their reasoning. Chapter 2 introduces students to the task of comparing data sets.

NAVIGATIONS SERIES

GRADES 6–8

NAVIGATING *through* DATA ANALYSIS

Chapter 2
Comparing Data Sets with Equal Numbers of Elements

Important Mathematical Ideas

In many situations, it is important to compare data from different groups—for example, crash-test data for different models of automobiles, the performance of groups of students on mathematics tests, or the reactions of men and those of women to particular kinds of advertising. Sometimes the data sets being compared have the same number of elements (equal *N*s), and sometimes they have different numbers of elements (unequal *N*s). This chapter deals with equal *N*s; unequal *N*s are addressed in chapter 3.

Deciding whether two data sets are "roughly similar" or "clearly different" depends on an understanding of the distribution for each data set. Data are the values assumed by a particular variable of interest. For example, in the activity involving TV watching (see fig. 1.3), the variable is the number of hours of TV watching in the previous week. The *distribution* of this variable is the values of the variable together with the frequency of each value. A graph of a data set is one way to display the distribution of the data; numerical summaries and measures of spread also provide useful information about a distribution. Deciding whether two data sets are similar or different consists essentially in deciding whether the distributions of the data sets are similar or different.

Spread and variability

Measurements are, by their very nature, variable. Since data are measurements of an attribute of the objects being studied, the data in a data

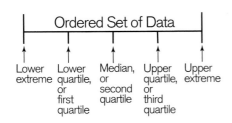

Ordered Set of Data

Lower extreme | Lower quartile, or first quartile | Median, or second quartile | Upper quartile, or third quartile | Upper extreme

Find, use, and interpret measures of center and spread, including mean and interquartile range.

set can always be expected to show variability. Understanding the differences in the variability, or spread, in two data sets is an important part of deciding whether the data sets are similar or different. The most common mathematical measure of variability is the *standard deviation*, but it is not included in the middle-grades curriculum. Kader (1999) discusses another measure that can be used—the *mean absolute difference*, or MAD.

The *range* (the difference between the greatest and least values in a data set) is one measure of the spread in a data set, but it is highly influenced by extreme values, so it may not be a good measure of variability for a particular data set. Another measure that is accessible to middle-grades students is the *quartile*. Quartiles are the three values that divide an ordered set of data into four equal-sized subsets. Of the data, 25 percent fall between two successive quartiles. Although the number of data elements between successive quartiles depends on the size of the data set, the percent of the data between successive quartiles is always 25 percent. The *interquartile range* is the difference between the upper and lower (or third and first) quartiles ($Q_3 - Q_1$). Half the data are in that range.

The easiest way to find the quartiles is to find the median of the data set and then to find the median of each half of the data. (The quartiles plus the least and greatest values are the five values needed to make a boxplot; boxplots are discussed in chapter 3; see also fig. 1.2.) If the quartiles for two data sets are almost the same, then the spread of the two data sets is likely to be quite similar.

Consider, for example, the data set 1, 2, 3, 4, 5, 6, 7, 8, 8, 9, 9. The median is 6. To find the first quartile, find the median of the lower half of the data. Here the lower half is 1, 2, 3, 4, 5, and the median of this half is 3. To find the third quartile, find the median of the upper half of the data. Here the upper half is 7, 8, 8, 9, 9, and the median of this half is 8. The quartiles, then, are 3, 5, and 8. The *first*, or *lower*, *quartile* is 3, the *second quartile*, or *median*, is 5, and the *third*, or *upper*, *quartile* is 8. The *interquartile range* is 8 − 3 = 5. When the number of data elements is odd, the median is the middle value of the ordered data; this value is *not* included in either the lower half or the upper half of the data. Note that the quartiles, like the median, may or may not be elements of the data set. If the number of elements in the data set—or in the halves of the data set—is even, then the quartiles are the means of the two center values and are usually not elements of the data set. So for the data set 1, 1, 2, 3, 4, 5, 6, 7, 8, 8, 9, 9, the median is 5.5, the first quartile is 2.5, and the third quartile is 8.

Shapes of distributions

Another way to describe the variability is by characterizing the shape of a distribution. The following are brief descriptions of some common shapes:

Mound-shaped distribution (bell curve). A mound-shaped distribution is generally symmetric, though perfect symmetry is very unusual. Mound-shaped distributions occur frequently; one example is the heights of all boys in a particular school (see fig. 2.1). The number of very short boys is about the same as the number of very tall boys, and the heights of the majority of the boys fall in the middle of the range. The values of the median and mean are close, and these values are close to the value represented by a vertical line through the peak of the mound. The quartiles are also clustered in the middle of the range.

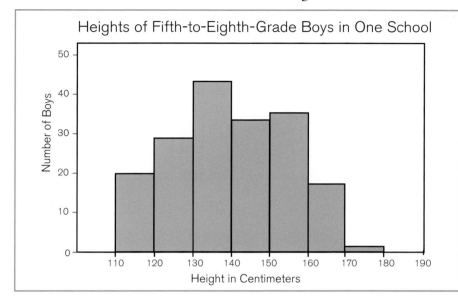

Heights of Fifth-to-Eighth-Grade Boys in One School

Fig. **2.1.**

A mound-shaped distribution

U-shaped distribution. A U-shaped distribution has two peaks, one at the upper extreme of the distribution and one at the lower extreme. See, for example, the U-shaped distribution in the duration in minutes of eruptions of Old Faithful in figure 2.2. The mean and median are close to the center of the distribution (i.e., close to the bottom of the U). The second quartile (or median) is close to the bottom of the U, the first quartile is close to the lower peak, and the third quartile is close to the upper peak.

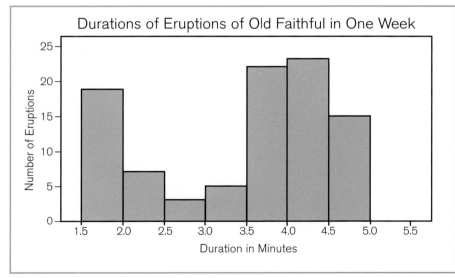

Durations of Eruptions of Old Faithful in One Week

Fig. **2.2.**

A U-shaped distribution

Rectangle-shaped (or *flat* or *uniform*) *distribution.* In a uniform distribution, each value is approximately equally likely to occur. For example, a bar graph of the number of births in the United States by month is uniform, as seen in figure 2.3. In an approximately uniform distribution, the mean and median are similar and close to the middle of the range. The quartiles are evenly spaced across the range.

Fig. **2.3.**

A uniform distribution

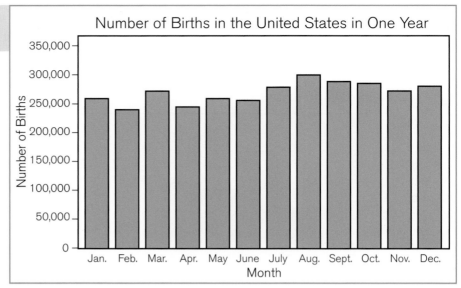

J-shaped (skewed left) or *backward-J–shaped (skewed right) distribution*. In skewed distributions, the data are bunched at one end of the range and stretched out at the other end. Skewed distributions are named according to the direction of the tail, not the location of the bulge. For example, a graph of the years in which roller coasters in the United States became operational is skewed left (see fig. 2.4) because more roller coasters have opened recently than opened earlier. In skewed distributions, the mean is pulled toward the tail, and the median is closer than the mean to the cluster of data at one extreme of the distribution. The median is often more representative of the data set than the mean. One of the quartiles is associated with the bunched data, and one quartile occurs toward the tail.

Fig. **2.4.**

A J-shaped distribution

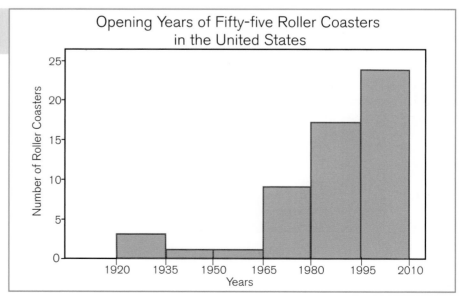

Extreme data values that are detached from the remainder of the data are called *outliers;* they usually receive special attention in statistical analysis. Statisticians often attempt to find explanations for outliers. They may represent legitimate values (e.g., one student in the class may be much taller than the other students), but they may also represent mistakes (e.g., an error in measuring or in recording a value). Outliers may also indicate that one of the data elements does not really belong to the population intended for study; in measuring the heights of the people in a second-grade class, for example, the height of the teacher is probably an outlier.

Reasoning about data

In the middle grades, students move from *additive reasoning* to *multiplicative reasoning*. In data analysis, additive reasoning involves comparing absolute frequencies (i.e., counts of the number of times particular data values occur), whereas multiplicative reasoning involves comparing relative frequencies (i.e., percents of the occurrence of particular data values). When two data sets have equal Ns, additive reasoning may be adequate. But additive reasoning does not apply to data sets with unequal Ns; in that case, multiplicative reasoning is required, as discussed in chapter 3. Data tables, line plots, stem plots, and histograms are all ways of displaying data so that these kinds of comparisons can be made.

Middle-grades students often approach the task of comparing data sets by focusing on specific values (e.g., the greatest values in the two sets) or by partitioning (e.g., determining the number of values in each set that are greater than a specific value). For example, to compare the performance of two golf teams of equal size, students might count the number of players on each team who scored below par or scored between 80 and 90. There is no need in this case to be concerned about the percent of players that scored below par; a decision about which team is better can be made simply by comparing counts.

Another common way for students to compare data sets is by comparing the "middle clump" of each data set. However, it is difficult for students to use these values for deeper analysis. For example, when asked to compare the heights of fifth graders with the heights of professional basketball players, fifth-grade students often identify the "typical" member of each group by the mode. However, when asked how much taller a typical basketball player is than a typical fifth grader, the students rarely find the difference between the two typical values they identified. Rather, they often focus on the difference between the greatest heights for the two groups.

What Might Students Already Know about These Ideas?

Students' approaches to comparing data sets are probably intuitive. The most common approach students are likely to take is to count the number of elements in particular intervals or above or below specific "cut points" and compare the counts. For situations in which the two sets have the same number of elements, this approach is often appropriate. It is essential, however, that in discussing these strategies, the teacher require the students to give a rationale for their choices of intervals or cut points. Some students will use multiplicative reasoning, which also is appropriate for the equal-Ns case, and it generalizes to the unequal-Ns case.

Each type of graph has unique characteristics, or conventions, that students should understand (see fig. 1.2). Curcio (2001) and NCSSM (1988) give ideas to help students understand and construct graphs.

The process of constructing a histogram is described in the essay "Making and Using Histograms," on the accompanying CD-ROM.

Dixon and Falba (1997) give examples of using data from the Internet to explore students' understanding of graphs.

As a preparation for students' comparing data sets with equal *N*s, you can assess their prerequisite knowledge about organizing data, choosing and making representations of those data, and justifying the choices of representation. The preassessment activity Students and Basketball Players asks students to identify the typical value in each of two sets and then determine the difference between the data in one set and the data in the other set. The intent is to assess whether students realize that the "typical difference" is the "difference of typicals." Many do not. Rather, they often compare specific values from the data sets, such as the maxima or the minima.

Students and Basketball Players

Goals

To assess students' ability to—

- read information from a data display;
- choose a representative statistic and justify the choice;
- use representative values to compare data sets.

Materials

- A copy of the blackline master "Students and Basketball Players" for each student

p. 88

Activity

To introduce this activity, you can ask the students what they know about the heights of professional basketball players—both male and female. You might ask the students how much taller their favorite basketball player is than the typical middle-grades student. (The typical height will, of course, differ for males and females and for sixth graders and eighth graders.)

Students should know what centimeters are, and they should have measured their heights in centimeters. If the students have reported the basketball players' heights in feet and inches, you should challenge them to convert the measurements to centimeters.

Distribute the copies of the activity sheet for the students to work on individually. You may want them to share their solutions with a partner before the whole-class discussion. The point of this assessment is to determine which students understand that the typical difference in heights is the difference in the typical heights and which students use other comparisons (e.g., a comparison of the greatest values in the two data sets).

Discussion

The students should be able to count the number of data elements in the stem plots (25 in each data set) and to explain that each element represents the height of a different person. Four students have heights of 152 cm, and eighteen basketball players have heights of at least 198 cm. These frequencies are simple counts, so students who cannot answer questions 1–3 may not be able to read the stem plots. You may need to review how a stem plot is constructed (see fig. 1.2).

Some students may answer questions 4 and 5 by identifying specific values: modes (152 cm and 205 cm), medians (151 cm and 203 cm), or means (approximately 149.7 cm and 201.6 cm). For each data set, the values for these numerical summaries are similar, so the students may not see a rationale for preferring one value over another. Other students may give ranges of values for questions 4 and 5. For example, they may say that the typical student is between 150 cm and 158 cm tall

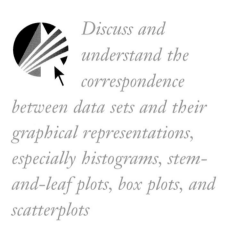

Discuss and understand the correspondence between data sets and their graphical representations, especially histograms, stem-and-leaf plots, box plots, and scatterplots

and the typical basketball player is between 200 cm and 207 cm tall; one argument for these choices is that they are the ends of the stems with the greatest number of leaves. Students often phrase this explanation, "Most of the data are between these values." It is true that more than half the data for the students are between those values. However, more data values for the basketball players are outside the interval from 200 to 207 than are in it. The use of the word *most*, therefore, is not accurate in this example. A correct phrasing of this idea is that the interval from 200 to 207 contains more values than the interval represented by any other stem. You have to decide whether the time is right to make this point explicit.

The intent of question 6 is to find out if the students will subtract whatever typical values they have identified for the two groups or whether they will focus on other values. Some students are likely to subtract the two extreme values (220 – 171). Some students may discount the value 171 cm in the student data because it is separated from the other data; these students may use 220 and 158 as the extreme values (220 – 158). Some students may also give a range of values for the difference, for example, 10 to 90, arguing that the difference from 170 (the stem with the greatest values of student data) to 180 (the stem with the least values of player data) is 10 and the difference from 130 (the stem with the least values of student data) to 220 (the stem with the greatest values of player data) is 90. Although this argument is correct, it is not very helpful for answering question 6.

Selected Instructional Activities

The following activities provide different contexts in which comparing data sets is important. They are in no particular order, but Classroom Climate involves a more complex data set, so it should probably not be the first task you present to the students. In each of the activities, the students are given two data sets with equal Ns. The students must then decide how to represent those data to facilitate the comparison of the data sets. For each activity, several representations are acceptable. In the follow-up to each activity, you may want the students to discuss the relative merits of the different representations.

In the activity Batteries, the data for Always Ready batteries span a noticeably larger interval than the data for Tough Cell batteries. This difference may cause some students to be unsure about which representations to use and may influence the inferences that students make. The students' justifications for their choices should help you identify the depth of their understanding of representations and numerical summaries.

Batteries

Goals

- Identify characteristics (e.g., mean, range, clusters) of data distributions
- Compare the characteristics of data in order to make decisions

Materials

p. 89

- A copy of the blackline master "Batteries" for each student
- Centimeter grid paper (available on the CD-ROM)
- A calculator or spreadsheet software

Activity

Distribute a sheet of grid paper and a copy of the activity sheet to each student. Call the students' attention to the data about the life of the two brands of batteries. Since the students did not actually collect the data themselves, it is important for them to think about how the data might have been gathered. Presumably, each battery was put into the same kind of device (e.g., flashlight, boom box), the device was turned on, and a record of the time was kept until the device quit working. The students need to understand that each datum represents the life of a single battery; once the device stops, the battery is dead and cannot be used again.

You can ask questions such as the following to lead the students to think about the conditions under which the device might have been used:

- Does ambient temperature affect the life of a battery?
- Should the temperature be constant?
- If so, does it matter if the constant temperature is very hot or very cold?
- Does humidity affect the life of a battery?
- Does the time of day affect the life of a battery?
- Does the kind of surface on which the device rests affect the life of a battery?

The activity has three parts. The first part (number 1 on the activity sheet) requires the students to think about which representations would be appropriate for displaying these data. The students might use line plots, stem plots, bar graphs, or histograms. One difficulty that students will face in graphing the data for Always Ready batteries is a large range (from 19 hours to 203 hours).

The second part of the activity (numbers 2 and 3 on the activity) requires the students to interpret the data; the representations students create should facilitate their interpretations. The last part (question 4) asks the students to think beyond the data and to use their interpretations to choose the brand of battery that has the longer life.

The students may analyze the data more completely if they work with partners or in small groups. As you observe the groups at work, help the students remain focused on the information in the data sets rather than on personal experiences. For example, some students may comment that they had a calculator "quit" during a test, so Mrs. Brewer should replace the batteries before each test. Although this concern may be important to that student, it is impossible from the data given to make any inferences about how Mrs. Brewer should handle calculator difficulties during tests.

Discussion

Have several students share their representations and their answers to question 2. Focus the attention of the class on how easy or difficult it is to interpret each of the graphical representations. For example, some students may create line plots, find the median for each data set, and then argue that the brand with the greater median is the better brand of battery. Other students may create graphs but base their argument solely on the computed values of the means, with the greater mean indicating the better brand of battery. In either case, you could ask how the graphs and the computed values show different information about the data.

The graph for Always Ready batteries is J-shaped (see the histogram in fig. 2.5), and the graph for Tough Cell batteries is mound-shaped (see the histogram in fig. 2.6). This difference in the shapes may help some students recognize the difference in the data sets. A solid argument can be made from comparing clusters of data in the two data sets. Line plots or stem plots of the data are other useful representations for revealing clusters (see the back-to-back stem plots in fig. 2.7).

The mode and median can be read quickly from the stem plot because the data have been ordered in it. Since each data set has forty entries, the median is located halfway between the twentieth and twenty-first entries. The mean has to be calculated; technology (e.g., a calculator or a spreadsheet) should be used for this task. The summary statistics are shown in table 2.1.

Friel and O'Connor (1999) offer ideas that help students compute quartiles.

Wilson and Krapfl (1995) discuss the use of technology in similar situations.

Table 2.1
Summary Statistics for the Battery Data

Statistic	Life of Always Ready Batteries in Hours	Life of Tough Cell Batteries in Hours
Mean	app. 98.0	app. 105.2
Median	105.0	104.5
Mode	115	110
Quartiles	87.5, 105, 113.5	97, 104.5, 110.5

Seventeen of the Always Ready batteries had a battery life greater than or equal to 110 hours, one battery had a life of 175 hours, and one had a life of 203 hours. However, seven of these batteries had a battery life less than 80 hours. Fourteen of the Tough Cell batteries had a battery life greater than or equal to 110 hours. The argument that students may find most compelling is that since the Tough Cell batteries are more consistent, Mrs. Brewer should use that brand.

Some students may draw histograms; the extreme values for the Always Ready batteries may make it problematic to determine the

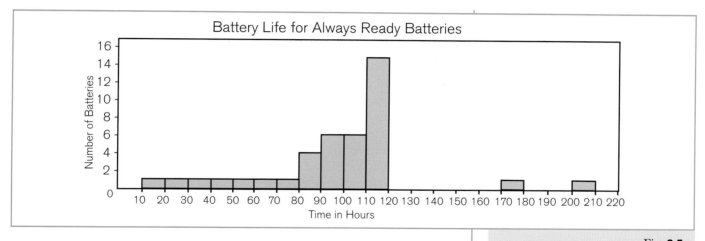

Fig. **2.5.**

A histogram of the data for Always Ready batteries

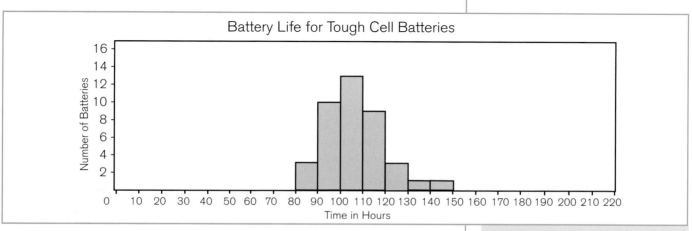

Fig. **2.6.**

A histogram of the data for Tough Cell batteries

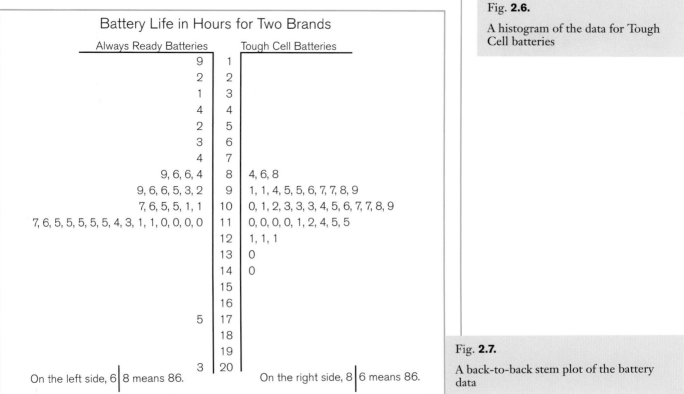

Fig. **2.7.**

A back-to-back stem plot of the battery data

Use observations about differences between two or more samples to make conjectures about the populations from which the samples were taken

appropriate intervals. It is important that the intervals be the same for the two data sets. Otherwise, it would be very difficult to compare the graphs. Computer software or graphing calculators would make the students' work much easier. With technology, the students can change the scale and the interval until they see a clear and reasonable picture of the distributions. Many students, however, may need to create one histogram on graph paper to have a clear understanding of how to interpret the electronic histograms.

Some students may say that they want "the battery that lasts 203 hours," which would indicate a confusion about what the data represent. Each datum is the number of hours that a particular battery (presumably chosen at random) lasted. That particular battery is dead, and it cannot be used again. There is no way to tell whether another battery, chosen at random, will be a "19 hour" battery or a "203 hour" battery. The point of analyzing these data is to find patterns that will apply to a new sample of batteries, without knowing how any particular battery will perform. The difference between interpreting a single data value (e.g., 203 hours) and interpreting an entire data set is not easy for students to grasp. Opportunities to work with several data sets are required for this idea to be internalized.

By changing the context, you can help the students understand the connection between the data and the context. For example, suppose you were a hiker and you needed a battery that lasts more than 110 hours for your cell phone. Which of these brands would you choose? A greater number of Always Ready batteries lasted more than 110 hours (13 for Always Ready, 10 for Tough Cell), so in this situation, the Always Ready brand might be the better choice.

Extensions

Possible extension activities include having students research actual brands of batteries (e.g., in *Consumer Reports* or on the Internet) or conduct experiments to determine how heat and cold affect battery life.

For further work on data sets with equal *N*s, see "Instructions for Users of Minitools" and Minitool 1, Using Case-Value Plots to Compare Data, on the CD-ROM. Four pairs of data sets for use with Minitool 1 have been supplied on the CD-ROM: Braking Distances of Cars, Yields from Cottonseed, Life Spans of Batteries, and Watermelon Juice. The context for each pair of data sets is explained in the "Instructions for Users of Minitools." Students use case-value plots to compare two data sets and answer a question about the comparison. Students can also enter their own data sets as files for use with this software. Directions for data entry can be found in the "Instructions for Users of Minitools."

In the activity Batteries, the students compare only two data sets. In the activity Stopping Distances, the students examine two pairs of stopping distances—one for 30 MPH and one for 60 MPH. After examining each pair, the students draw a general conclusion about which model of car seems safer; that is, the students are asked to make two conclusions, each based on data, and then to draw an inference on the basis of the two conclusions. By attending to the language that the students use for these two kinds of activities, you can determine how clearly they understand the differences in the tasks.

McClain, Cobb, and Gravemeijer (2000) describe how seventh-grade students used Minitool 1 in an experimental statistics unit.

Stopping Distances

p. 90

Goals

- Represent data appropriately
- Use the characteristics of data to make decisions

Materials

- A copy of the blackline master "Stopping Distances" for each student
- Centimeter grid paper, available on the CD-ROM

Activity

To introduce this activity, you can raise the issue of vehicle safety by asking, for instance, "What are some things that car buyers consider as they make their choices of model?" The students may identify several considerations: cost, desirability or popularity, engine power, size, safety, intended uses, or reliability. Focus on safety by asking, "What are some of the aspects of safety that a buyer might be concerned about?" The students may identify several issues, for example, braking distance, number and placement of air bags, and type of bumper. The data in this activity deal with braking distances.

Distribute an activity sheet and grid paper to each student. The students should work with a partner or in small groups to analyze the braking-distance data on the activity sheet and complete the three tasks. The first two involve comparing the breaking distances of two models of automobiles at two speeds independently. The third task requires the students to combine the first two analyses to arrive at a single conclusion. Emphasize that each group must prepare an argument that everyone in the group agrees with. This requirement is intended to encourage the students to negotiate their positions and be ready to justify the group's choice.

Discussion

Since both the data sets are small, some students may calculate the numerical summaries for each set (see table 2.2) or create line plots (see figs. 2.8–2.11). Although the mean and median suggest that the large sedan stops in shorter distances, the greater range and the lack of a mode for stopping distances at 30 MPH for the large sedan suggest that the buyer could not be confident that the stopping distance of a large

Table 2.2
Summary Statistics for Stopping Distances at Two Speeds for Two Automobiles

| Statistic | Stopping Distances in Feet | | | |
| | Small Sedan | | Large Sedan | |
	30 MPH	60 MPH	30 MPH	60 MPH
Mean	64.4	147.2	62.5	143.8
Median	65.5	147.5	60.5	137.5
Mode	65, 67, 68	147, 154	none	137
Range	11	30	28	38

sedan would fall within desired limits. The stopping distances are more consistent for the small sedan at 30 MPH. Although three of the large sedans stopped within 50 to 55 feet at 30 MPH, three large sedans also required 70 feet or more to stop at that speed. The same information—as well as the individual data points—can be seen from a back-to-back stem-and-leaf plot (see figs. 2.12 and 2.13). One conclusion that students might draw from these data is that more data are needed to make a good decision.

Fig. **2.8.**

A line plot of the stopping distances for the small sedan at 30 MPH

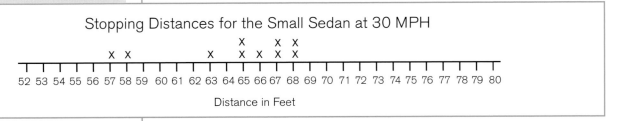

Fig. **2.9.**

A line plot of the stopping distances for the large sedan at 30 MPH

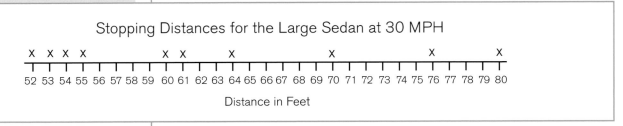

Fig. **2.10.**

A line plot of the stopping distances for the small sedan at 60 MPH

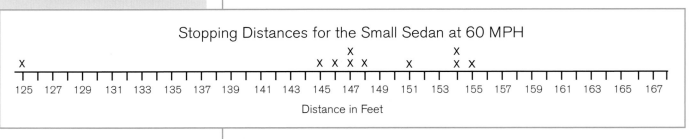

Fig. **2.11.**

A line plot of the stopping distances for the large sedan at 60 MPH

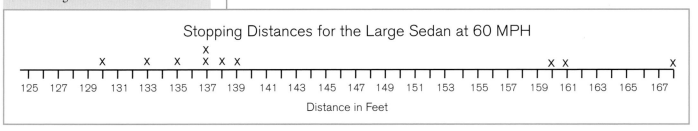

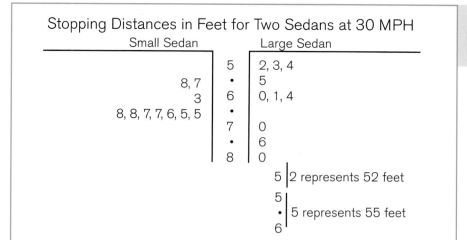

Stopping Distances in Feet for Two Sedans at 30 MPH

Fig. **2.12.**

Back-to-back stem plots of stopping distances at 30 MPH for two models of sedans

Small Sedan		Large Sedan
	5	2, 3, 4
8, 7	•	5
3	6	0, 1, 4
8, 8, 7, 7, 6, 5, 5	•	
	7	0
	•	6
	8	0

5 | 2 represents 52 feet

5 |
• | 5 represents 55 feet
6 |

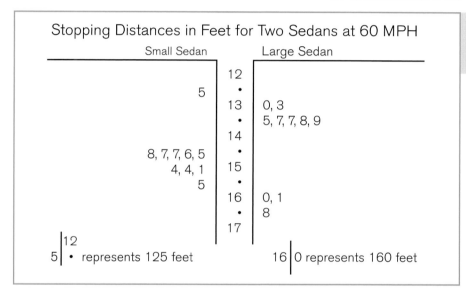

Stopping Distances in Feet for Two Sedans at 60 MPH

Fig. **2.13.**

Back-to-back stem plots of stopping distances at 60 MPH for two models of sedans

Small Sedan		Large Sedan
	12	
5	•	
	13	0, 3
	•	5, 7, 7, 8, 9
	14	
8, 7, 7, 6, 5	•	
4, 4, 1	15	
5	•	
	16	0, 1
	•	8
	17	

5 |12
5| • represents 125 feet   16 |0 represents 160 feet

On the basis of these data, a car buyer would probably conclude that the large sedan has the safer stopping distances at 60 MPH because seven out of ten of the stopping distances for the large cars are less than nine out of ten of those for the small sedans. However, the three longest braking distances were recorded by large sedans. The students might say they want more data before drawing any conclusions.

The following activity, Classroom Climate, asks students to examine a relatively complex data set. Some students may treat all the data for a classroom as a single set of data, but other students may insist on separating the data according to time of day. Having the students discuss their reasons for these two approaches will help them understand the complexity of the data.

Classroom Climate

Goals

- Compare multiple data sets of equal size
- Make decisions based on characteristics of data

Materials

- A copy of the blackline master "Classroom Climate" for each student
- Half-centimeter grid paper (available on the CD-ROM)

Activity

Ask the students if different parts of their school building are always the same temperature. If they reply that they are not, ask, "What might cause differences in temperature?" Among the causes that the students may identify are the following: the kinds of heating or air conditioning units, the number of vents, the number of windows, the heights of the ceilings, and differences in the various stories of the building. The data in the activity focus on the differences between the first and second floors of the building and on its orientation, which influences the amount of sunlight coming in through the windows.

Distribute an activity sheet and grid paper to each student. The students should work in small groups. Several data representations (e.g., graphs, measures of center) must be made, and the work will go more quickly if several students work together. The major part of the work, however, is making reasonable arguments about which rooms are colder and by how much.

The activity involves making two analyses, comparisons of data about the first and second floors and of data about the north-facing and south-facing rooms. Each analysis is a comparison of data for two pairs of classes. Some students may want to refine the comparisons by separating the temperatures by time as well as by location. Doing so makes the analysis much more complex; students who wish to attempt this degree of complexity should be encouraged to do a simpler analysis first.

Discussion

Distribute grid paper and a copy of "Classroom Climate" to each student. Before analyzing the temperature data, the students will have to decide whether to treat the data in any column as one data set or as four data sets (one for each of the different times). It is possible, for example, that the data at 9 A.M. might be different from the data at 3 P.M. This difference would not be noticed if the data in each column were treated as a single data set. Some students may begin by locating the minimum and maximum temperatures in each column and by calculating measures of central tendency. They may or may not include the mode. The results are shown in table 2.3.

Table 2.3
Statistics for Data on Classroom Temperatures

	All Data in a Column Treated as a Single Data Set			
Statistic	Room 203	Room 103	Room 204	Room 104
Mean (°F)	69.5	68.3	67.9	66.9
Median (°F)	69.5	68	68	67.5
Minimum (°F)	63	62	61	60
Maximum (°F)	75	74	73	72
Range (°F)	12	12	12	12

	Data Separated by Times			
Statistic	6 A.M. Data			
Mean (°F)	64.6	63.4	62.8	61.0
Median (°F)	64	63	63	61
Minimum (°F)	63	62	61	60
Maximum (°F)	66	65	64	62
Range (°F)	3	3	3	2
Statistic	9 A.M. Data			
Mean (°F)	68.0	67.2	66.8	65.8
Median (°F)	68	68	67	66
Minimum (°F)	66	65	65	65
Maximum (°F)	70	68	68	67
Range (°F)	4	3	3	2
Statistic	Noon Data			
Mean (°F)	72.0	70.2	70.2	70.2
Median (°F)	72	70	70	71
Minimum (°F)	69	67	68	68
Maximum (°F)	75	73	73	71
Range (°F)	6	6	5	3
Statistic	3 P.M. Data			
Mean (°F)	73.2	72.4	71.6	70.4
Median (°F)	74	72	72	70
Minimum (°F)	71	71	70	69
Maximum (°F)	74	74	73	72
Range (°F)	3	3	3	3

It is clear from the data that the second-floor rooms are warmer than the first-floor rooms. The challenge comes in estimating how much warmer each floor is. One estimate is calculated by taking the difference of the means. For the data considered as a single set, the means of the two second-floor rooms are 69.5°F and 67.9°F; the means for the two first-floor rooms are 68.3°F and 66.9°F. The averages of these pairs of means are 68.7°F for the second-floor rooms and 67.6°F for the first-floor rooms; the difference is 1.1°F. A similar computation on the medians yields average medians of 68.8°F for the second-floor rooms and 67.8°F for the first-floor rooms, with a difference of 1°F. The students can be challenged to argue why one statistic might be more representative of the actual temperatures than the other is. Some students may argue that other computed statistics are more appropriate ones to compare. What is important here is whether they can make a logical argument for one statistic over another.

VanLeuvan (1997) provides an example of the use of similar kinds of graphs.

It is also clear from the data that the south-facing rooms are warmer than the north-facing rooms. For the data considered as a single set, the means of the two south-facing rooms are 69.5°F and 68.3°F; the means for the two north-facing rooms are 67.9°F and 66.9°F. The averages of these pairs of means are 68.9°F and 67.4°F; the difference is 1.5°F. A similar computation on the medians yields average medians of 68.8°F for the south-facing rooms and 67.8°F for the north-facing rooms, with a difference of 1.0°F.

Since only five data values were recorded for each room at each time of day, making graphs for each room of the five temperatures at each of the different times of day is not very profitable. However, graphing the means by room at the four times of day (see fig. 2.14) reveals some similar patterns. For the problem at hand, however, looking at the summary statistics seems to be the better way of analyzing the data.

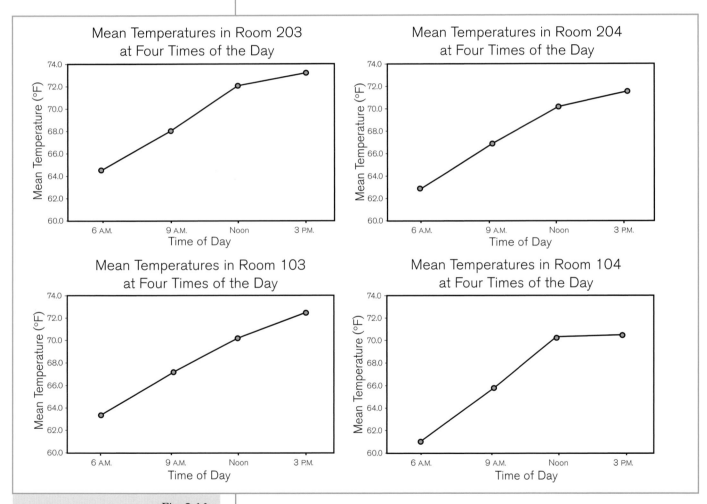

Fig. 2.14.

Graphs of mean temperatures for four classrooms at four times of the day

Conclusion

All the activities in this chapter involve data sets with equal *N*s. When the *N*s are unequal, the reasoning required becomes more complicated. Chapter 3 addresses the issues related to comparing data sets with unequal *N*s.

NAVIGATING *through* DATA ANALYSIS

Chapter 3
Comparing Data Sets with Unequal Numbers of Elements

Important Mathematical Ideas

This chapter focuses on ways to compare data sets when the numbers of elements in the data sets are not the same (unequal Ns). The mathematics needed to compare data sets with unequal Ns is the same whether the Ns are nearly the same or dramatically different. Students, however, approach these situations differently. For example, if they are comparing the performance of two golf teams, one with twenty-five players and one with twenty-seven players, many students will suggest that they "simply ignore" two of the scores from the larger team so that the remaining data sets have the same number of players. That is, those students try to reduce this situation to an equal-Ns case. The process they suggest for choosing the two scores to ignore is often biased; for example, they may propose ignoring the players with the greatest and least scores. The examples in this chapter deal only with cases in which the Ns are remarkably different.

When the Ns are dramatically different—for example, when the performance of seventh graders in a school having one seventh-grade class is compared with the performance of seventh graders in a school having five seventh-grade classes, students pay attention to the fact that the Ns are unequal and realize that it is not feasible to adjust the numbers of data elements. Most students are more likely to see the need to compare the proportions of elements in common intervals—that is, to use multiplicative reasoning. The development of multiplicative reasoning is one goal of middle-grades mathematics instruction, and data-comparison situations with dramatically different Ns elicit this kind of reasoning as students realize that to compare some data sets, they must use relative frequencies rather than absolute frequencies.

Relative frequencies

For a single data set, an absolute-frequency graph and a relative-frequency graph have the same shape, except that the graph may appear to be stretched or shrunk vertically because of a possible difference in scale on the vertical axis (see the examples in figs. 3.1 and 3.2). The only other difference is the label on the vertical scale. It is important that students recognize the similarity of these two kinds of graphs.

Fig. 3.1.

An absolute-frequency graph of the lengths of cats

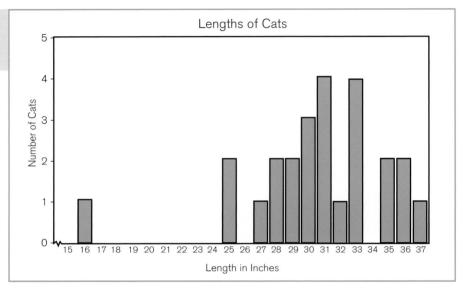

Fig. 3.2.

A relative-frequency graph of the lengths of cats

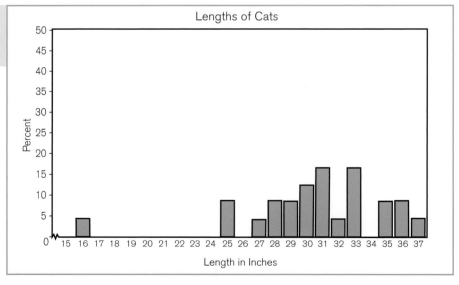

The *relative frequency* is the frequency of occurrence of a given data value *relative to,* or *as a fraction of,* the total number of elements in a data set. This value can be written as a ratio that is easily converted to a percent. For example, if six out of thirty people like vanilla ice cream, the relative frequency is 6/30, or 20 percent, of the data set.

Similarly, for two data sets with equal Ns, the absolute-frequency and relative-frequency graphs have the same shape (again, except for possible scale differences), since the relative frequencies for the values in both data sets are computed by dividing by the same number, that is, the common number of data elements.

For two data sets having unequal Ns, however, the absolute-frequency and relative-frequency graphs may show very different patterns. It is important to relate these patterns to the reasons for comparing the two data sets with unequal Ns. Understanding the reasons for comparing the data sets will demonstrate the need for relative-frequency graphs. Consider, for example, the graphs in figures 3.3 and 3.4. The data, from the activity A Matter of Opinion, are from responses to two different survey questions.

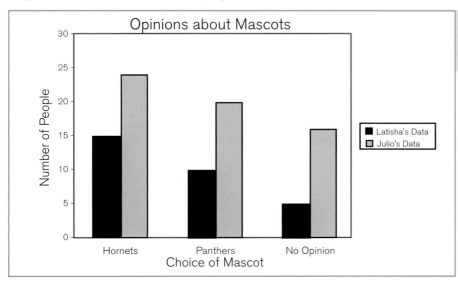

Fig. **3.3.**

An absolute-frequency (bar) graph of opinions

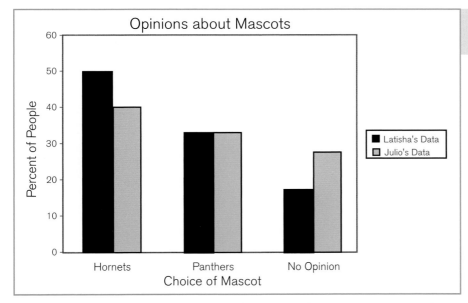

Fig. **3.4.**

A relative-frequency graph of opinions

In the absolute-frequency graph (fig. 3.3), Julio's question gives the impression of generating more support for all options, which is due in part simply to the greater number of people surveyed by Julio. The relative-frequency graph (in fig. 3.4) presents a different picture. Latisha's question generated a greater range of intensity; the support for Hornets was as great as the support for the other two options combined. Julio's question generated more nearly equal support for the three options. Latisha's data would be more persuasive than Julio's in arguing that the mascot should be Hornets. The relative support for Panthers is the same for both questions. Julio's data would be more persuasive than Latisha's in arguing that there is no overwhelming support in the community for either choice. The position of the community is not clear, however, unless the data are displayed in a graph like the relative-frequency graph in figure 3.4.

In order to make sound decisions on the basis of data with unequal Ns, students need to learn how to create relative-frequency graphs. Intuitively, it would seem easier for them to learn this skill by making both absolute- and relative-frequency graphs for a single set of data. Doing so, however, would probably obscure the need for the relative-frequency graph, since the shapes of those two graphs would be the same. It is only in comparing data sets with different Ns that the need for and advantages of the relative-frequency graph become apparent.

When data are numeric, relative-frequency data can be used to compute the quartiles. Quartiles separate the data into four subsets with 25 percent of the data in each subset; computing quartiles from relative frequencies reinforces this idea. To compute quartiles, it is necessary to know only the percent of data points associated with each value. As a step toward finding the quartiles from relative-frequency data, it is often helpful to create the cumulative percents for the ordered data. For example, for the cereal data in table 3.1, 9.1 percent of the boxes had zero grams of sugar per serving, and 16.9 percent of the boxes had three grams of sugar per serving. The cumulative-percent column shows that 15.6 percent of the boxes had two grams or less of sugar per serving and that 32.5 percent of the boxes had three grams or less of sugar per serving, so the first quartile is three grams. That is, the first value that 25 percent of the data are less than or equal to is three grams. Using similar reasoning, you can see that the second quartile, or the median, is 7 (since the 50% mark is reached there), and the third quartile is 11 (since the 75% mark is reached there).

Of the data, 25 percent are at or below the lower quartile and 75 percent are above the lower quartile. Likewise, 75 percent are at or below the upper quartile and 25 percent are above the upper quartile.

Table 3.1
The Amount of Sugar per Serving of Cereal

Grams of Sugar per Serving		Percent of Boxes	Cumulative Percent
Minimum →	**0**	**9.1**	**9.1** ← **0% of boxes**
	1	2.6	11.7
	2	3.9	15.6
Quartile 1 →	**3**	**16.9**	**32.5** ← **25% of boxes**
	4	1.3	33.8
	5	6.5	40.3
	6	9.1	49.4
Median →	**7**	**5.2**	**54.6** ← **50% of boxes**
	8	6.5	61.1
	9	5.2	66.3
	10	6.5	72.8
Quartile 3 →	**11**	**6.5**	**79.3** ← **75% of boxes**
	12	9.1	88.4
	13	5.2	93.6
	14	3.9	97.5
Maximum →	**15**	**2.6**	**100.1** ← **100% of boxes**

Box plots

Another graphical representation that offers many of the same benefits as relative-frequency graphs is a *box-and-whiskers plot*, or *box plot*. A box plot is made from five values: the minimum, the first (or lower) quartile, the median (or second quartile), the third (or upper) quartile, and the maximum. The "box" of this display denotes the middle 50 percent of the data; the difference between the two ends of the box is called the *interquartile range*. To make the box, simply locate the lines that denote the middle 50 percent of the data and make a box as shown in figure 3.5. Then connect the ends of the box to the extremes with line segments; these segments are the "whiskers." For the cereal data in table 3.1, the five values are 0, 3, 7, 11, and 15; the box plot of these data is shown in figure 3.5.

Detailed directions for making box plots are given in Friel and O'Connor (1999).

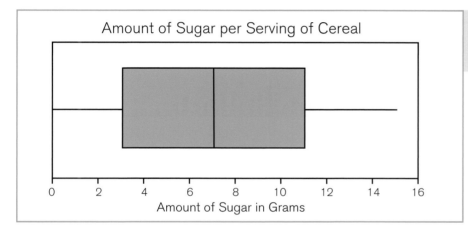

Amount of Sugar per Serving of Cereal

Amount of Sugar in Grams

Fig. **3.5.**

A box plot of the data on the amount of sugar per serving of cereal

Box plots can be "stacked" on a single graph with the same horizontal scale, as shown in figure 3.6. The five values needed to create box plots represent "cut points" for the same percents of data values, so box plots are useful for comparing data sets when the numbers of data are different, especially if the numbers are very different. This point is illustrated in the activity Migraines: Box Plots. Basing a display on percents avoids the necessity of explicitly accounting for differences in numbers of data elements.

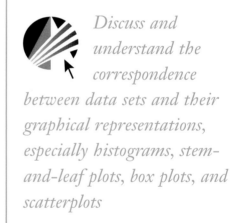

Discuss and understand the correspondence between data sets and their graphical representations, especially histograms, stem-and-leaf plots, box plots, and scatterplots

Find, use, and interpret measures of center and spread, including mean and interquartile range

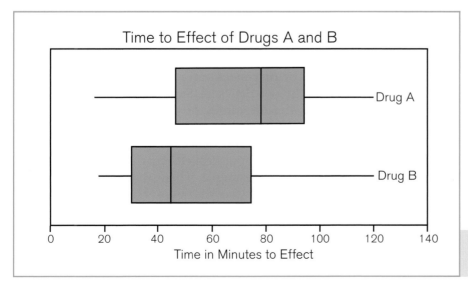

Time to Effect of Drugs A and B

Drug A

Drug B

Time in Minutes to Effect

Fig. **3.6.**

Stacked box plots

It may be easier for students to understand a box plot when it is superimposed on a line plot or a dot plot (see fig. 3.7). To compute quartiles, students often write the data in order and then "count off" the quartiles. This approach may cause confusion about what scale to use for the box plot. Students tend to use the entire list of ranked data elements as the scale and may become confused by duplicate values or values omitted from the list. Creating a box plot directly on a line or dot plot helps students overcome these difficulties, since the scale of the line plot is clear, with neither duplicate nor omitted values.

Fig. **3.7.**

A box plot superimposed on a dot plot

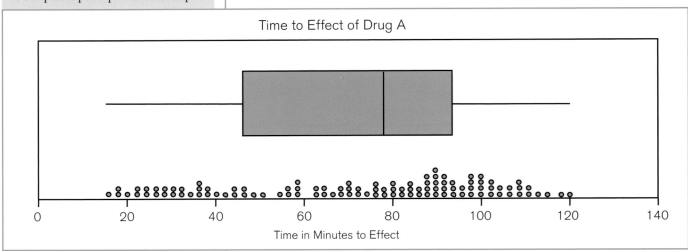

Time to Effect of Drug A

Time in Minutes to Effect

Multiplicative reasoning

The use of relative-frequency graphs and box plots supports multiplicative reasoning about data, since both tools display the distribution of percents of data independent of the actual number of elements. Indeed, if the only information a person has about a data set is a relative-frequency graph or a box plot, it is impossible for him or her to determine how many elements are in the data set.

In data analysis, multiplicative reasoning involves the comparison of relative frequencies within a data set as well as among data sets. In many situations, it is impossible, or at least very expensive, to have the same number of elements in two data sets. Deciding whether two data distributions with different Ns are similar requires a comparison of rates or ratios. Using simple counts—that is, additive reasoning—is not adequate. Comparing quartiles is an example of multiplicative reasoning, since it involves comparing concentrations of data (i.e., ratios of numbers of elements) in the two data sets. Such comparisons are illustrated directly in the activities Migraines: Histograms and Migraines: Box Plots.

Numerical summaries—that is, mean, median, and mode—can also be used to compare two data sets without explicit regard to the number of data points in either data set. For example, when students are asked which of two drugs is faster acting (see Migraines: Histograms and Migraines: Box Plots), some students find all three numerical summaries and identify the drug that "won," according to each summary. If all three summaries indicate that the same drug is the faster one, these students are likely to say that drug is better, independent of how the distributions of the data compare. It is important, however, to relate these numerical summaries to the data distributions. The shape of the data—including clusters, holes, and outliers—and the spread of the data—as revealed by the quartiles—are also important to consider.

What Might Students Already Know about These Ideas?

In order to use relative-frequency graphs and quartiles in comparing data sets, students need to understand how to read such displays of data. Assessing students' understanding is the goal of the following activity, Cereal. Students' responses will also reveal if they understand percents well enough to use multiplicative reasoning effectively. A thorough understanding of percent as a rate is necessary for the development of such reasoning.

Cereal

Goal

To assess students'—
- ability to read information from a relative-frequency graph;
- understanding that it is not possible to read the number of data elements from a relative-frequency graph;
- ability to identify quartiles from a relative-frequency graph.

Materials

- A copy of the blackline master "Cereal" for each student

p. 92

Activity

Set up the activity by having the students who ate cereal for breakfast identify themselves and then asking those students to share the names of the cereals that they ate. Ask, "How do the cereals you mentioned differ in food value?" The students may suggest several categories, such as the amount of sugar, fiber, protein, or various vitamins. This activity focuses on sugar content, so ask how sugar in cereals is measured. Help the students understand that it is usually measured in the number of grams per serving.

Distribute the copies of the activity sheet, and ask the students to work on it individually. Then bring the class together for a whole-group discussion of their answers. If the students have difficulty with this activity, use the data in the activities in chapter 2 to make relative-frequency graphs, and use those graphs to give students additional experiences.

Discussion

The first two questions are read-beyond-the-data questions; the students have to read specific information from the graph and then perform computations using that information. Question 3 is not answerable from the data; a relative-frequency graph does not show the actual number of data elements. Questions 4 and 5 focus attention on the middle of the distribution. The students should begin to understand that relative frequencies provide enough information to compute the median. Question 6 extends this awareness to the computation of the other quartiles.

Selected Instructional Activities

The activities in this chapter offer students opportunities to develop skills in making relative-frequency graphs and box plots and then to use these representations to compare data sets with unequal *N*s. The last two activities are built around a common context—namely, the effectiveness of medications. You may want to ask your students to investigate how drugs are tested for effectiveness. You or some of your

Find, use, and interpret measures of center and spread, including mean and interquartile range

Software may be useful in these activities. Kader and Perry (1994) give examples of similar uses of software.

students may know medical professionals who can be invited to give the class an explanation of the effects of prescription medications on patients.

The data presented in the next activity, A Matter of Opinion, are intended to illustrate clearly the differences between absolute-frequency graphs and relative-frequency graphs. In addition, the activity affords an opportunity to discuss how differences in the wording of survey questions might influence the results of a survey.

A Matter of Opinion

Goals

- Demonstrate skill in constructing absolute-frequency and relative-frequency graphs
- Identify similarities and differences between the two kinds of graphs
- Use a relative-frequency graph to make decisions

Materials

- A copy of the blackline master "A Matter of Opinion" for each student
- Half-centimenter grid paper, available on the CD-ROM

Activity

Ask the students to think about how they might go about surveying the community to obtain people's opinions. They may suggest phone surveys, mail surveys, or face-to-face surveys. You can discuss the relative advantages and disadvantages of these methods, but that is not the central point of this activity. Discuss why it is sometimes important to know the opinion of the community. As a specific example, ask, "Why might it be important to know what the community thinks of the proposed name of a new school's mascot?" The students may suggest that if the community found the name offensive, it might not support the school. This activity deals with the analysis of data generated by two different questions, each of which was intended to gauge the opinion of the community. The students can use the data generated by the responses to these questions to examine how the wording of a question might influence the way people respond to it.

Distribute the activity sheets and grid paper. The students can work individually or in pairs. Working in pairs might promote discussion of the best ways to interpret the data. You may want to interrupt the students after they have answered question 5 to discuss the absolute-frequency graph before they move on to creating a relative-frequency graph. If you hold such a discussion, use it as an opportunity to assess how the students are thinking about the two sets of data before they examine the relative frequencies. Lead a discussion of students' answers to the remaining questions to compare the differences in students' reasoning about the relative-frequency graphs.

Discussion

This activity is intended to help students understand that a relative-frequency graph is a better data display than an absolute-frequency graph to compare two data sets with unequal Ns. The activity will also reveal differences in the ways that students reason about these two kinds of graphs. Categorical data were intentionally selected so that computing numerical summaries, such as mean or median, would not be possible, thereby keeping students' attention focused on the relationship among frequencies and the relationships among relative frequencies.

p. 93

The design and use of surveys is not addressed in this book; however, a good discussion and several activities for students (Zawojewski 1991, pp. 4–26) are available on the CD-ROM.

See Curcio (2001) for a discussion of how to make the graphs.

Questions 3, 4. and 5 focus students' attention on a direct comparison of the data from the two surveys. The students who are already thinking multiplicatively may express some discomfort at this point in the activity, saying that Julio's data show more support than Latisha's for both the mascots. The discomfort comes from an intuitive recognition that simply comparing the frequencies ignores important information—namely, the relationship between the frequencies and the total number of elements in the data set. Other students may be completely comfortable with the direct comparison, arguing that Julio's data show stronger support for both names. Their explanations of their position will reveal their use of additive rather than multiplicative reasoning.

Question 7 asks the students to compare the absolute-frequency graph with the relative-frequency graph. In the absolute-frequency graph (see fig. 3.3), each of the bars for Julio's data is taller than the corresponding bar for Latisha's data, but in the relative-frequency graph (see fig. 3.4), only one of the bars for Julio's data is taller. Latisha's question seems to have elicited relatively more support for Hornets than Julio's question did. The two questions seem to have elicited relatively the same support for Panthers, and Julio's question seems to have elicited relatively more "no opinion" responses than Latisha's question did. The relative-frequency graph displays these patterns clearly. Questions 8 and 9 are specifically designed to help students see the differences in how people reacted to the two questions. Latisha's data show stronger relative support for Hornets than Julio's data show, but the relative support for Panthers seems to be the same in both sets of data.

The next two activities deal with a single context—the effectiveness of drugs. Analyzing these data using both histograms and box plots permits an extensive interrogation of data in a single context, an exercise that will help students understand that data analysis is a sophisticated process.

Migraines: Histograms

Goals

- Use histograms and relative-frequency histograms to analyze data
- Extract information from a histogram and a relative-frequency histogram

Materials

- A copy of the blackline master "Migraines: Histograms" for each student
- Calculators (optional)
- Half-centimeter grid paper, available on the CD-ROM

Activity

To introduce this activity, ask the students what they know about how new drugs are tested. Once a new drug is ready to be tested in humans, it is likely to be tested with only a small number of patients. If it is being compared with a drug already in use, much more data will be available for the approved drug. Data for the two drugs will, therefore, have dramatically different Ns. For this activity, assume that drug A is a traditional drug that has been approved and used for some time and that drug B is the new drug. This scenario explains the large discrepancy between the numbers of data points in the two data sets.

This activity demonstrates to students the difference in the utility of an absolute-frequency histogram and a relative-frequency histogram. The relative-frequency histogram allows a direct comparison from graphs of data sets that would otherwise be difficult to compare because they contain unequal Ns. If the students do not know how to make a histogram, instruct them in the construction of histograms either as part of this activity or as a prelude to it.

The scenario presented on the blackline master indicates that each patient records her or his own information. You might want to raise the issue of self-reporting and ask the students whether they think this method might affect the accuracy of the information. Distribute grid paper and a set of activity sheets to each student. Calculators, if available, will help the students compute the relative frequencies. The students should work with a partner or in a small group to encourage rich discussions.

Discussion

Some students are likely to compare these distributions by choosing a "cut point" and counting the number of values in each distribution above or below that cut point. Suppose, for example, that you were interested in knowing how many patients in each group received relief within sixty minutes. Thirty-six patients took drug A and received relief in sixty minutes or less, but only thirty-one patients who took drug B received relief in sixty minutes or less. The students might conclude that drug A is faster acting. This direct, additive comparison is faulty, however, because the numbers of patients using the drugs differ dramatically. A better strategy is to find the percent of the patients who

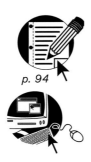

p. 94

An essay on constructing histograms, "Making and Using Histograms," has been included on the CD-ROM.

McClain, Cobb, and Gravemeijer (2000) provide more insight about how students reason about data.

received relief within sixty minutes. About 34 percent of the patients taking drug A and about 66 percent of those taking drug B received relief within that time. The percent for drug B is almost twice that for drug A.

Doing analysis with cut points, however, has some disadvantages. The most important is that often no rationale can be given for the use of any particular cut point. Choosing sixty minutes, for example, seems intuitively reasonable, since it is an hour and people might expect to receive relief within an hour. But is sixty minutes a better (or worse) cut point than sixty-four minutes or fifty-three minutes or even seventy-seven minutes? Without knowing something about the physiology of migraine headaches, we probably cannot make a medically and statistically appropriate choice. A cut point, then, is likely to be completely arbitrary, at least from a mathematical point of view, and different cut points can often lead to different conclusions. Another disadvantage is that for each cut point selected, the percent of data values above or below it must be recomputed. These disadvantages can be overcome somewhat with the use of a relative-frequency histogram. The following is an example of how one teacher led a discussion of this activity that helped her students reason about the proportions of data in particular intervals.

Teacher: In a test conducted at a major hospital, people who suffered from migraine headaches were given one of two drugs. Drug A has been on the market for several years and has been found to provide relief to many people who suffer from migraine headaches. Drug B is a newer drug that has the potential to give faster relief to more people. The drugs were given to people with migraine headaches, and the subjects were asked to record the amount of time it took them to get relief after taking the drug. We have results from 106 people who took drug A and 47 people who took drug B.

Doug: Why didn't they do the same number of people for each drug? That would be more fair.

Teacher: Why do you think they don't have the same number of results for each drug?

Montez: Because they probably gave it to the same number of people but just got the results after, say, a month, and that's how many people had headaches in that time.

Teacher: Very nice, Montez. This brings out a very important fact about comparing data sets, and that is that they are often not the same size. Will that make a difference in how you conduct your analysis? You might want to think about that.

The students looked at the data and tried to answer question 1 on the activity sheet. The teacher brought them together for a brief period of sharing.

Kyle: We found out how many patients in each group got relief in less than forty minutes. We found that 22 patients who took drug A got relief in less than forty minutes and only 20 who took drug B, so we chose drug A.

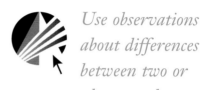

Use observations about differences between two or more samples to make conjectures about the populations from which the samples were taken

Andrea: We did the same thing, but we used thirty minutes and we got 12 with drug A and only 9 with drug B.

Shade: Wow, so either way, drug A is better!

At this point, the students were focused only on the data in the range of zero to thirty or forty minutes. Although their direct, additive comparisons did tell them how many patients received relief in a short time, the students were ignoring an important aspect of the analysis. They were not reasoning about the fraction, or the percent, of the patients that fell in the range from zero to thirty or forty minutes. The teacher suggested that they make the histogram and relative-frequency histogram "to see if those representations change your mind." After students have constructed the displays, it is important to discuss the different features of the graphs, which this teacher did:

Teacher: What conclusions can you draw about the effectiveness of the two drugs by comparing the relative-frequency histograms that you have just completed?

Kyra: Well, it looks like drug A has the highest percentage of people at the high end of the graph.

Teacher: Can someone else say what they think Kyra is saying? Jose?

Jose: She is saying that if you look at the graph of drug A, then it has its highest bars, which mean the largest percentage of patients, at the high end, meaning it took more time to get relief.

Teacher: So, on the basis of what Kyra and Jose said, would you recommend drug A?

Dante: No, you want drug B because you want your tall bars, or your high percentages, to be low, where it takes less minutes to get relief.

Teacher: Yeah, but I thought drug A had some high bars in the low numbers. I mean, look at twenty to forty minutes. They have a bunch of people here.

Dante: Yeah, but it's a very low percentage of all the people that took drug A, so you really don't have a very good chance of getting relief quick with drug A like you do with drug B.

The importance of this discussion is that it helped the students understand what information can be extracted from a comparison of the relative-frequency histograms. Discussions of activities, such as this conversation, are an essential part of helping students make sense of and improve their reasoning.

The following activity, Migraines: Box Plots, extends the analysis of the drug data. Learning how to construct a box plot is one of the goals of this activity, but it is more important that students understand how the box plots help them interpret the data. Box plots are particularly valuable for interpreting the spread of data.

Migraines: Box Plots

Goals

- Practice creating box plots
- Extract information from box plots and use that information to make decisions

Materials

- A copy of the blackline master "Migraines: Box Plots" for each student.

p. 97

Activity

The context and data for this activity are the same as for Migraines: Histograms. The focus here is on making box plots, extracting information from the box plots, and then drawing conclusions. Distribute a set of activity sheets to each student. The students may make faster progress if they work with a partner.

Discussion

The range of fifty percent of the data can be determined by looking at the median and one of the extremes. For example, for drug A, the lower 50 percent of the data are between sixteen minutes and seventy-eight minutes; for drug B, the lower 50 percent of the data are between eighteen minutes and forty-four minutes. These statistics present a rather compelling argument in favor of drug B, since half the patients received relief in forty-four minutes or less.

Locating the quartiles helps make a stronger argument. The quartiles are the upper limit of the lower 25 percent and the lower limit of the upper 25 percent of the data. In conjunction with the median and the extremes, the quartiles divide the data into four groups, each of which contains 25 percent of the data. These values form the five-point summary (see table 3.2).

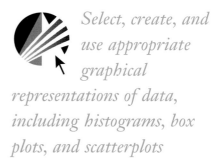

Select, create, and use appropriate graphical representations of data, including histograms, box plots, and scatterplots

Table 3.2
The Five-Point Summary for the Drug Data

	Least Value	First Quartile	Median	Third Quartile	Greatest Value
Drug A	16 min	46 min	78 min	94 min	120 min
Drug B	18 min	30 min	44 min	74 min	120 min

For drug A, 25 percent of the data are forty-six minutes or less, whereas for drug B, 50 percent of the data are forty-four minutes or less. For drug A, 50 percent of the data are seventy-eight minutes or less, whereas for drug B, 75 percent of the data are seventy-four minutes or less. Drug B seems clearly to provide faster relief for a greater percent of patients. We can imagine the same teacher who led the classroom discussion for the previous activity continuing with a discussion of this activity with box plots:

Many students have trouble reading a box plot. They interpret it to mean that the farther apart the bars are the more data the interval includes. It is important to emphasize the idea that 25 percent, or one-fourth, of the data fall in each interval. This idea can be reinforced by writing "25%" at the top of each interval.

It is important to stress in different ways what can be inferred about the distribution of the data from the width of the intervals. The interval width does not indicate the frequency of the data in the interval. Each interval contains 25 percent of the data. The interval width indicates how clustered or spread out the data are.

McClain, Cobb, and Gravemeijer (2000) describe how seventh-grade students used Minitool 2 in an experimental statistics unit.

Teacher:	OK, using the graphs you just made that show the five lines on the axis, what conclusions can you draw about which drug is faster?
Sarah:	I think that drug B is faster.
Teacher:	What is the basis of your decision, Sarah?
Sarah:	OK. Well, there are two groups in drug B below forty-four minutes and only one group in drug A below forty-six.
Teacher:	So what does that tell me if I get migraine headaches?
Juan:	I know! Half the people who took drug B got relief in forty-four minutes or less, but only one-fourth of the people who took drug A got relief in forty-six minutes or less.
Sharika:	There's another way to look at it.
Teacher:	OK, Sharika, what is that?
Sharika:	Well, you could also say that 50 percent of the people who took drug A had to keep hurting for at least seventy-eight minutes, but only 25 percent of the people who took drug B had to hurt that long.
Teacher:	Hmm, so who is correct? Juan or Sharika?
Students:	Both.
Teacher:	But how can that be?
Jamie:	'Cause it just depends on how you want to say it. There's lots of ways that are right as long as you understand the graph.
Teacher:	OK, let's look at the two graphs for the two drugs. Notice where the line for the median is located in each box. What does that tell us about the data?
Megan:	In drug A, the second 25 percent of the data are more spread out than in drug B.
Teacher:	That's right. The space between the end of the box and the median line tells us the range of that quarter of the data.
Kyle:	Yeah, so if we know that 25 percent of the data are in each interval, then the ones that are close together have the data bunched up, so that should be where the clump or cluster of the data is.

Extensions

It is interesting to give students a box plot and ask them what the data might look like. Obviously, many different distributions can generate the same box plot, so this task could lead to a lively discussion about which data distribution is the most believable.

For further work using dot plots to compare data sets with unequal Ns, see "Instructions for Users of Minitools" and Minitool 2 on the CD-ROM. The sets of data to be used with Minitool 2 Using Dot

Plots to Compare Data, that are already on the CD include AIDS Treatments, Ambulance Response Times, Cardiovascular Disease, Cholesterol Levels, Heights of Corn, Flu Shots, Heart Rates, Migraine Treatments, Speed Trap, and Weight Changes. Students can use dot plots to compare two different data sets in different contexts (e.g., data about several times for two different ambulance companies to reach destinations) and answer a question about the comparison. The different contexts are explained in "Instructions for Users of Minitools." Students can also enter their own data sets as files to use with this software. Data-entry directions can be found in "Instructions for Users of Minitools."

Conclusion

All the activities in this chapter involve data sets with unequal Ns. The contexts in which such comparisons occur are common in the real world. Typically, these contexts are also complex, so students need time and experience to develop the kind of reasoning required to analyze such complex data. Sophisticated reasoning is also required in situations in which multiple measurements are made on each case—that is, in situations involving bivariate data. Students can explore bivariate data in the activities in chapter 4.

NAVIGATING *through* DATA ANALYSIS

Chapter 4
Exploring Bivariate Data

Important Mathematical Ideas

Bivariate data

When two characteristics are measured for each object in a sample or population, the resulting data are *bivariate*. Bivariate data are paired data that give information about two different characteristics (or attributes or variables) of the elements in a sample or population. For example, data showing daily high and low temperatures for one month in Chicago are bivariate data; the objects, or cases, are the days of the month, and the two characteristics are daily high temperature and daily low temperature. However, the variables represented by bivariate data do not have to be numerical. Data showing the gender and political-party affiliation of state representatives are bivariate data with two categorical variables. A list of the ratings and running times for current movies is an example of bivariate data with one categorical variable and one numerical variable.

In examining bivariate data sets, one of the questions to consider is whether the data *covary*—that is, whether changes in one variable are related to changes in the other variable. If a relationship exists, we may want to know whether it is direct or inverse and whether it is strong or weak. For example, there is a direct relationship between the population of a state and the number of members of the U.S. House of Representatives from the state, since House membership is proportional to the population of the state. Some variation is found in the populations of states with, say, six U.S. representatives, so the relationship between population and number of House members is not perfect, although it is fairly strong. However, there is no relationship between the population of a state and the number of members in the U.S. Senate, since each state has two senators, independent of population. An example of an inverse relationship is the relationship between the travel time and the average speed for driving between two cities; as the average speed increases, the time decreases. So we can say that number of members in the House and state population covary directly and that travel time and average speed covary inversely.

Scatterplots and prediction

When the variables are both numerical, any relationship that may exist between them can be described mathematically. A *scatterplot* is a graph drawn to represent bivariate, numerical data. Examining a scatterplot gives students an opportunity to visualize whether a relationship exists between the two variables. Relationships between variables can be either linear or nonlinear and either direct or inverse, or there may be no relationship at all.

Students need opportunities to explore many different data sets that demonstrate many different kinds of relationships (or no relationship). An important part of interrogating bivariate data is deciding whether a relationship exists between the two variables, so students need to have opportunities to explore data without being directed to look for a particular kind of relationship between the variables.

When two variables are related linearly, the scatterplot of their data will approximate a straight line; the stronger the relationship, the more closely the data will approximate a straight line. When it exists, this line is called the *line of fit*; it describes the general trend of the data. When a scatterplot suggests that a linear relationship exists, students can lay a piece of uncooked spaghetti or line segments drawn on a transparency on the scatterplot to help them approximate a line of fit. Students can use their approximate lines of fit to see trends and make reasonable predictions. A good approximation of a line of fit is determined by sophisticated statistical calculations. A discussion of different approximations (formed by, e.g., moving the uncooked spaghetti up or down or by rotating it) can help students investigate how a line of fit could be determined.

One benefit of using scatterplots is that they are visual representations of data in which general trends can be examined. Students have opportunities to interpolate and predict values not present in a display by estimating where the values will fall, given an observed general trend. For example, there have been several years during which the Olympic Games did not occur. Students can examine the existing data to speculate about what the winning times or distances might have been if the Olympics had been held in those years. They can also predict the winning times or distances for future events; that is, an approximate line of fit can be used to predict values beyond the range of existing data. As the saying goes, "time will tell" whether such predictions are accurate.

Matsumoto (1981; see the CD-ROM) and Rubink and Taube (1999) suggest ways to incorporate scatterplots in instruction.

Select, create, and use appropriate graphical representations of data, including histograms, box plots, and scatterplots

Students should be aware that slight differences in a line of fit may result in large differences in predictions, particularly when the predictions go far beyond the range of the existing data. For example, in the preassessment activity Reading a Scatterplot, students are asked to predict the number of telephone area codes a country would have for a given population if it followed the U.S. system. Typically, within the range of the given data (i.e., 5 to 20 million people) students' interpolations are fairly accurate. However, predictions of area codes for a population beyond the data—for example, a state with 40 million people—often vary widely from one student to the next. Having students display their lines of fit on the same scatterplot will illustrate why those predictions are so different. The lines of fit might vary only slightly within the range of the given data, but as those lines are extended, the differences among them become magnified.

Students should also realize that predictions from data have limitations. A classic example involves the winning times of the men's Olympic 200-meter dash over the years (see the activity Exploring Relationships). The winning time decreases as the year increases, and the relationship seems to be inverse and approximately linear. If so, the line of fit would be a straight line with a negative slope. But if such a line is extended, it eventually crosses the horizontal axis. A point on the part of the line that is below the horizontal axis would represent a year in which the winning time is less than zero, and the "winner" would have to have finished the race before he started! Clearly this situation is impossible, but it illustrates that care is needed in extending a line of fit too far beyond the existing data. A relationship may not be very strong, so the line of fit may not come very close to the data, or the relationship may be nonlinear, so a straight line of fit may not be very useful.

Students should come to understand that relationships suggested by the data as displayed in scatterplots are not necessarily causal. It is important to distinguish between variables that simply covary "by accident" (usually because of some other underlying cause) and variables that illustrate a causal relationship. For example, a scatterplot showing the number of telephones and the number of teachers for various cities in Illinois suggests that the number of teachers increases whenever the number of telephones increases. This relationship does not mean that increasing the number of telephones in a city will increase the demand for or supply of teachers in the city. Nor does an increase in the number of teachers in a city cause a city to install more telephones. In this example, an external factor—namely, increases in the population of the city, may be a cause for both increases. An increase in a city's population might result in both more telephones being installed and more teachers being hired. These two variables are "correlated," but neither is a cause of the other. In contrast, the longer a pot of water is heated on a stove, the hotter the water gets (at least up to its boiling point). Longer heating times can be interpreted as "causing" higher temperatures. Understanding the differences between correlation and causation does not happen quickly or easily, but addressing these issues informally is a reasonable goal for middle-grades mathematics instruction.

Scheaffer (2000) discusses how statistics in general fits into the mathematics curriculum of the future.

What Might Students Already Know about These Ideas?

During the middle grades, students should develop skill in plotting points on a grid; they should know the difference between the x-coordinate and the y-coordinate of a point. They also need to know how to name the coordinates of a particular point. This skill is essential for accurate interpretation of the data in a scatterplot. Beyond interpreting individual points, however, students need to develop skill in identifying trends in the data represented in a scatterplot. The following activity, Reading a Scatterplot, will let you assess the level of the students' understanding and skill.

Reading a Scatterplot

Goals

To assess students'—

- skill in identifying the two values associated with each point in a scatterplot;
- ability to interpret trends in data in a scatterplot;
- ability to use trends to make predictions.

Materials

- A copy of the blackline master "Reading a Scatterplot" for each student.

p. 100

Activity

This activity explores the relationship between the populations of states and the number of telephone area codes in the states. Begin by asking, "Why do phone companies add area codes to telephone numbers?" The students should be able to figure out that the reason is that the phone companies begin to run out of seven-digit phone numbers to assign to customers. Ask, "Why do phone companies begin to run out of phone numbers to use?" The students may know that the shortage is a result of a greater use of regular phones, cellular phones, fax machines, and communication lines for data or computers. Ask, "Why is there a greater demand for phone lines?" The students may say that the demand is due to more businesses and to an increasing population. The activity focuses on population as a variable that is correlated with the number of area codes in the states.

Distribute a copy of the activity sheet to each student. The students should work with a partner to help enrich their interpretation of the data. As you observe the pairs working, encourage them to give clear explanations of the reasoning behind their answers.

Discussion

The numbers of area codes in the states are discrete data—for example, there cannot be 3.5 areas codes in a state—so the graph on the activity sheet may seem somewhat unusual to students. Questions 1 and 2 familiarize students with reading data points on a scatterplot by asking them to read information directly from the graph. For question 1, the students need to count the number of dots above the value 5 on the horizontal axis. For question 2, the students have to interpret the two values associated with point *A*.

Questions 3, 4, and 5 ask the students to make predictions or interpolations from the data. Questions 3 and 5 ask for an interpolation within the range of values of the data, and question 4 asks for predictions that go beyond the range of the data. The students may find an approximate line of fit to make their predictions. Through oral discussion or the students' written explanations, you should be able to discern whether the students have some sense of the linear trend of the data. That is, their predictions should not be just wild guesses; they should

Discuss and understand the correspondence between data sets and their graphical representations, especially histograms, stem-and-leaf plots, box plots, and scatterplots

be supported by such arguments as "As the population increases by x million people, the number of area codes seems to increase by y."

Question 6 assesses the students' ability to describe the relationship between the two variables. The point of the question is not to elicit an exact answer. Rather, the responses will allow you to assess whether the students can use their observations logically to formulate a description of the relationship that makes sense (both to you and to other students). For example, to begin to make sense of the trend, some students may focus on how to find a representative value for the stack of dots displayed above each value on the horizontal axis (number of area codes). One strategy is to pick the value in the middle of these dots; another strategy is to pick a value in the greatest concentration of the stack of dots. More information about students' responses to these questions can be found in Mooney (2002).

Selected Instructional Activities

The following activities afford students several opportunities to create scatterplots and then to identify and describe trends in the data. That is, the activities help students explore whether two variables covary and if so, what the nature of that covariation is. Learning how to make scatterplots is one goal of these activities, but it is not the main one. It is more important for students to learn to interrogate bivariate data, whether presented in tables or in graphs, just as they learned to interrogate univariate data from the activities in the first three chapters.

Students must go beyond simply reading data from a scatterplot; they need to learn how to represent bivariate data. They do so in the activity Congress and Pizza, but they may not elect to use a scatterplot. If they do not, then discussing the students' other representations can help you learn how the students are thinking about bivariate data.

Congress and Pizza

Goals

- Represent bivariate data
- Determine the relationship between two variables

Materials

- A copy of the blackline master "Congress and Pizza" for each student
- Half-centimeter grid paper, available on the CD-ROM

p. 102

Activity

A single context is used for the three activities Congress and Pizza; People, Congress, and Pizza; and Predicting. This context is the relationship among the populations of the states, the number of pizza restaurants in the states, and the number of U.S. representatives for the states. Population is the underlying factor in the second and third variables, and the activities are intended to lead students to that conclusion. Data for only forty states are presented in the activity Congress and Pizza; the students are asked to predict or interpolate the values for the other ten states in the activity Predicting.

House (2001) provides more information about congressional apportionment.

Set the stage for this activity by asking, "Which two states would you expect to have the greatest number of pizza restaurants?" "Why?" The students may identify high-population states, such as California, New York, or Illinois; they may identify states in some specific region of the country, such as the Southwest; or they may identify states, such as New York, Massachusetts, and Illinois, that are perceived as having large populations of Italian-Americans. Follow up by asking, "Which two states would you expect to have the fewest pizza restaurants?" "Why?" The students should identify states that have characteristics that are the "opposite" of the characteristics of the states they identified previously.

Then turn to consideration of the numbers of U.S. representatives. Ask, "Which two states would you expect to have the most U.S. representatives?" "Which two states would you expect to have the fewest U.S. representatives?" "Why?" The students should know that the number of representatives is proportional to the population, except that each state is guaranteed at least one representative, even if its population is very small.

Distribute grid paper and an activity sheet to each student. The students can work either individually or in pairs, depending on your assessment of their level of skill in creating scatterplots. If the students need experience in drawing scatterplots, you may want them to draw the graphs individually. When they start to describe the relationship between the variables, however, have the students work with a partner to encourage discussion about the relationship.

Discussion

You may want to start the discussion of this activity by asking the students about situations in which a relationship might intuitively be clear to them. For example, you could ask what has changed about them as they have become older. Focus on responses that show a direct relationship with getting older—for example, getting taller, staying up later, receiving a greater allowance. This discussion will help students look at how a change in one variable results in a change in another variable.

The students will probably use a variety of ways to represent the data in the activity. Have them share several of these representations, and for each, ask what information can be read from the display. For example, some students may consider ordering the data twice: once according to the number of representatives and a second time according to the number of pizza restaurants. Others may order the data on the basis of only one variable. They must understand, however, that the two values for each state must be kept together; the variables should not be ordered separately. Bar graphs and double-bar graphs could be used. Someone may use a scatterplot, but if not, you may have to introduce that graph as an alternative display of the data. One benefit of a scatterplot is that it offers a highly visual representation of the trends in the data. A scatterplot allows the students to see the linearity of the relationship between the two variables, and this feature might not be seen in a table or a double-bar graph. Also, the scatterplot keeps the values of the variables for each data pair together. Bar graphs and some table arrangements treat the data as two separate and independent variables.

The students may initially be unsure of how to describe the relationship they observe. If the difficulty continues, ask, "If you connected the dots from left to right, what would you get?" (The "shape" would be approximately a straight line that runs from the lower left to the upper right of the graph.) You may have to introduce, or at least remind the students of, correct terminology by explaining that a straight line represents a *linear relationship*. The precise use of terminology, however, is not the main point. Different language is acceptable as long as the explanation shows an understanding of the direct, linear relationship between the variables.

This activity gives students a chance to explore two related variables, neither of which is the cause of the other. The next activity, People, Congress, and Pizza, extends the investigation by exploring the underlying cause of the relationship—namely, population.

Make conjectures about possible relationships between two characteristics of a sample on the basis of scatterplots of the data and approximate lines of fit

People, Congress, and Pizza

Goal

- Understand that relationships are not necessarily causal

Materials

- A copy of the blackline master "People, Congress, and Pizza" for each student

- Half-centimeter grid paper, available on the CD-ROM

- Graphing software for creating scatterplots (optional)

p. 103

Activity

The students will need to have access to a scatterplot of the number of U.S. representatives versus the number of pizza restaurants in the forty specified states. If they did not do so as part of the activity Congress and Pizza, you should have them create the scatterplot before you begin this activity. If graphing software is available, you may want the students to use it to create the graph.

Begin this activity by asking, "In the last activity, you discovered that the number of pizza restaurants and the number of U.S. representatives are positively related. What do you think might explain this relationship?" It is likely that some students will suggest that a state's population is important. Some students may suggest that the location of the state is important. For example, warm states have more pizza restaurants, since people order pizza to avoid heating up their houses by cooking dinner. In the states where the winters are severe, pizza restaurants may experience difficulty during the winter in delivering pizza while it is still hot. You might point out that these explanations could account for differences in the numbers of pizza restaurants in northern and southern states and ask the students if their explanations also account for the differences in the numbers of representatives.

Other questions you might ask are the following:

- Do states with more U.S. representatives have more political power to get pizza restaurants in their states than states with fewer U.S. representatives?

- Do states with more pizza restaurants have the ability to get more U.S. representatives elected than states with fewer pizza restaurants?

- How are the numbers of U.S. representatives determined?

The purpose of this discussion is to help the students realize that population is strongly related to the other two variables.

Distribute the activity sheets and grid paper to the students. Since the data are fairly complex, the students should work in pairs or small groups so that the work of constructing the graphs can be shared.

The number of U.S. representatives and the number of pizza restaurants are related, but neither causes the other. The underlying mechanism is population. Therefore, all three variables are related, even though only two variables at a time can be graphed on a scatterplot.

Discussion

This activity serves to connect mathematics and social studies by exploring how the numbers of U.S. representatives for each state are determined. You may want to coordinate the use of this activity with the social studies teacher.

Ask some students to display the scatterplots they have made. You could post the scatterplots on a wall or on a bulletin board so that the students can see them. The discussion of the scatterplots should focus on how tightly the dots are grouped around an invisible straight line. The scatterplots for population and U.S. representatives will illustrate the tightest grouping; that is, the data for those two variables will better approximate a straight line than the data for the other two pairs of variables will. Some students may say this scatterplot shows the strongest relationship. Be sure that the focus is on the tightness of the clustering of the data.

Begin the discussion of question 3 by asking, "What do you notice about the population of states with one U.S. representative?" (The populations are quite variable because each state is guaranteed a minimum of one representative.) Follow up by asking, "Is there as much variability in the populations of states with six representatives?" (No, since the ratios of the numbers of representatives that exceed 1 are proportional to the population.) Then let the students explain the relationships that they identified. Be alert to language that illustrates an understanding of the positive linear relationships between the number of U.S. representatives and the population and between the number of pizza restaurants and the population.

Extensions

Have the students look up other data that might be correlated with the number of representatives—for example, the number of schools in a state, the number of hospitals in a state, the number of newspapers in a state, and so on.

Ask the students to imagine what a scatterplot for state population and the number of U.S. senators would look like. No relationship exists, since each state, independent of population, has two senators. If the students graphed population on the horizontal axis and number of senators on the vertical axis, all the dots would lie on the horizontal line $y = 2$.

The next activity, Predicting, extends the investigation of this context even further. The students are asked to predict or interpolate the data for the ten states whose data were omitted from the list. The students can then compare their predicted or interpolated values with the actual values.

Formulate questions, design studies, and collect data about a characteristic shared by two populations or different characteristics within one population

Predicting

Goal

- Use scatterplots and approximate lines of fit to predict or interpolate data values

Materials

- A copy of the blackline master "Predicting" for each student
- Uncooked spaghetti or line segments cut from overhead-transparency copies of the blackline master "Line Segments for Approximating Lines of Fit"
- The scatterplots the students made for the activities Congress and Pizza and People, Congress, and Pizza

pp. 104, 106

Activity

This activity continues the exploration begun in the previous two activities. Ask the students if they noticed that data from some states were not included in the earlier activities. Ask which states were missing.

The students will need copies of the scatterplots they created in the previous activities. If these activities have not already been completed by the students, you may want the students to use technology to create the necessary scatterplots.

Distribute to each student a copy of the activity sheet and a few pieces of spaghetti or several transparency copies of the line segments for approximating lines of fit. The students should work in pairs or small groups to encourage discussion. Depending on your students' mathematical backgrounds, you may want them to approximate a line of fit visually, to calculate an equation for the approximate line of fit, or to use technology to determine the actual line of fit. Whichever way you choose, you may need to discuss how the students might determine approximate lines of fit for the data in Congress and Pizza and in People, Congress, and Pizza. They could then use those lines to predict or interpolate the missing values.

Discussion

Have the students share their results for the ten states and compare their answers. The results for relatively small values for numbers of U.S. representatives and state populations are more likely to be closer than the results for relatively large values of numbers of U.S. representatives and state populations. Slight differences in the slope of the approximate line of fit will show a greater difference at greater values than at lesser values. Have the students place on the same scatterplot several line segments approximating lines of fit with different slopes and compare the differences in the predicted or interpolated values that result from the different approximate lines of fit.

For each state, the students must use two scatterplots to predict or interpolate the missing data. For each value that the students interpolate or predict, conduct a discussion of which scatterplots are the most useful. If the students use different scatterplots, have them compare the accuracy of the different results. Focus on the strength of the relationship (i.e., the closeness of the points to a line of fit) in a scatterplot as a factor that might influence the accuracy. Because the relationship between the population and the number of U.S. representatives is the strongest of the three relationships, the scatterplot of those data would be expected to yield the most accurate population predictions.

Discuss that predictions are just that—predictions; they are not expected to match the actual values exactly. When discussing factors that might influence the differences between the predicted or interpolated values and the actual values, consider the idea of outliers. Are some data points extremely different from other data points? If they were removed, how would the line of fit be affected? (An outlier might "pull" the line of fit toward that point, influencing the slope of the line.)

In the previous four activities, students have made scatterplots and used them to predict and interpolate data. The next activity, Exploring Relationships, gives students opportunities to explore relationships (or the lack of a relationship) in four different contexts.

Exploring Relationships

Goals

- Explore relationships in data
- Make predictions and explore the limitations of those predictions

Materials

- Copies of the blackline masters "Olympic Gold Times," "Population Trends," and "Time Is Money?" for each student
- Almanacs or access to the Internet
- Half-centimeter grid paper, available on the CD-ROM
- Uncooked spaghetti or line segments cut from overhead-transparency copies of the blackline master "Line Segments for Approximating Lines of Fit"

pp. 107, 108, 109

Activity

The three data sets in this activity can be explored independently and in any order. The goal is to give the students opportunities to use scatterplots to represent bivariate data and then to interpolate or predict values.

If graphing software or graphing calculators are available, the students could use it to draw scatterplots. One of the challenges in managing this use of technology is that the students have to enter the data before the graphs can be drawn. You may want the students to use the data files that are available on the CD-ROM, or you could have each of several students create a different master file. The students could then share the files to avoid having every student spend time entering the data.

Discussion

These data sets show different relationships:

- Olympic Gold-Medal Times for the 200-Meter Dash—negative linear relationship
- U.S. Population 1790–2000—nonlinear positive relationship
- Running Times and Gross Receipts of Movies—no relationship

The 200-meter-dash data give you a chance to discuss the limitations of predictions in some situations. When predicting athletic records, for instance, human capabilities must be taken into consideration. At some point, it will be impossible for a person to run a 200-meter dash as fast as predicted. You could ask the students to find data from other Olympic events and explore the relationships of those data to the year of the event. Is there a limitation on how far a discus can be thrown or how high a person can pole vault?

In the U.S. population data, the students can see a nonlinear relationship between two variables. Again, you can discuss the limitations of the display in predicting the population in future years. For example, if the population trend were to continue indefinitely, the population would become so large that unsustainable overcrowding would result. The movie data present variables with no relationship.

Data sets from which the students could create scatterplots can be found in almanacs and U.S. Census data on the World Wide Web (www.census.gov), among other sources. Students can also collect data on their own to use to make scatterplots.

Extension

For further work in using scatterplots to explore relationships between data sets, see "Instructions for Users of Minitools" and minitool 3 on the CD-ROM. Sets of data to be used with minitool 3, Using Sctterplots to Explore Relationships between Two Variables, on the CD include Alcohol and Reaction Time; Education and Salary—Women, Education and Salary—Men; Bone Mineral Density—American Women, Bone Mineral Density—Japanese Women; Traffic—1992, Traffic—2000; Brushing and Plaque—Brand A, Brushing and Plaque—Brand B; and Speed Reading—Program 1, Speed Reading—Program 2. Each data context is explained in the "Instructions for Users of Mini-tools." Students use scatterplots to represent these data sets and answer questions about the relationships observed.

Conclusion

Making and interpreting scatterplots involve sophisticated mathematics content that builds on an understanding of simpler data-analysis concepts and skill in creating simpler representations. The activities in this book have illustrated one way to make the journey, by interrogating a data set, comparing data sets, and seeing relationships between two variables. The completion of this journey prepares students to move forward to a deeper understanding of data analysis in high school.

Minitool 3 on the CD-ROM allows teachers and students to extend the activity.

NAVIGATING *through* DATA ANALYSIS

Looking Back and Looking Ahead

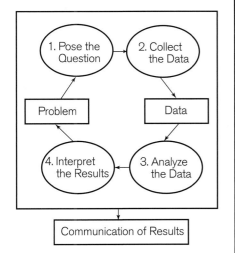

In the middle grades, work in data analysis begins with examining a single data set and moves through comparing two data sets to analyzing bivariate data. At the beginning of this journey, students can use additive reasoning, but by the end, they need facility with multiplicative reasoning. Along the way, students develop increasingly sophisticated language and ways of representing data that help them not only communicate their reasoning but also engage in reasoning. Emphasizing data analysis as a process (see the illustration) supports students' learning of data analysis.

The two main tools of data analysis that are available to middle-grades students are graphs and numerical summaries. Students need to understand how to read and interpret the information in graphs, how to make graphs that are correct and that communicate the important aspects of data, and how to compute and interpret measures of center and simple measures of spread. Some attention needs to be given to developing the skills necessary for graph making and computing measures of center and spread, but the main emphasis of instruction should be on understanding how data provide information about an interesting question. The question might be of practical interest—involving the effectiveness of medications, for example—or it might be somewhat more frivolous—for example, relating the numbers of U.S. representatives and the numbers of pizza restaurants in states. In every instance, however, the question should motivate students to explore important mathematical ideas.

Much of the reasoning that middle-grades students use in data analysis is informal. More-formal topics, such as creating lines of best fit or testing hypotheses, will wait until high school or beyond, when more-sophisticated mathematical tools are available to students. Drawing a box plot or describing the shape of a graph may be a new strategy for middle-grades students. These strategies build on the knowledge that students gain in elementary school. The challenge for teachers is to help students choose strategies carefully and then communicate clearly how those strategies support the conclusions that they make about the data and the context.

Doing multiple tasks in a single context in which more and more data are explored can help students develop connections among the ideas of data analysis. The time to relief from medication is a context that might be used in this way. Initially, students could investigate what a doctor might be able to tell patients about how fast they might receive relief from a traditional migraine medication. The students could examine data for one drug, determine how to represent those data so that patients could easily understand the information, and explore what the various measures of center or spread might communicate about the speed of the effect of that drug. Then they could examine data on an experimental drug. There might be an equal number of patients in the studies—(equal Ns)—or an unequal number—(unequal Ns). By comparing the data sets, the students could decide whether the experimental drug appears to give faster relief than the traditional drug. Finally, the students might study more-detailed questions about the speedier drug. Is it equally fast for all ages of patients? Is it equally fast for patients who have only recently begun to have migraines and for patients who have had migraines for several years? These questions would require multiple measurements from the patients, and the analysis would require such tools as scatterplots.

Technology can be a useful support for middle-grades students' learning of data analysis, although it should not take the place of direct, hands-on work with data (e.g., making frequency tables of data or constructing graphs on paper). Graphing software and displays from graphing calculators can help students make different kinds of graphs of data quickly and easily, but the software will not determine which of the possible types of graphs are appropriate and which are not. In a particular context, students should discuss which graphs make sense to use with a data set and which do not. They should also discuss what each acceptable graph reveals about the data and the context.

Middle-grades teachers have the challenge of helping students make the transition from the relatively unsophisticated reasoning appropriate in elementary school children to the more formal reasoning that is developed in high school. Providing a wide range of experiences with different kinds of data sets is necessary to this process. Across the middle-grades curriculum, students should examine large and small data sets, compare data sets with equal and unequal Ns, and study data sets that show multiple measurements on the same cases. Dealing with all these ideas requires careful planning and sequencing of instruction. Connecting data analysis with content outside mathematics—science and social studies, for example—can help teachers make effective use of the instructional time available and help students develop important connections between mathematics and other disciplines. Underlying all instruction in data analysis, however, is the idea that is most important for students to learn: that data analysis is a process that helps make sense of a situation.

NAVIGATIONS
SERIES

GRADES 6–8

NAVIGATING
through
DATA ANALYSIS

Blackline Masters and Solutions

Lengths of Cats

Name _____

A group of students has been investigating information about their pets. Several students have cats. They decided to collect some information about each of the cats. One set of data they collected was the lengths of the cats measured from the tip of the nose to the tip of the tail. Here is a bar graph showing the information they found:

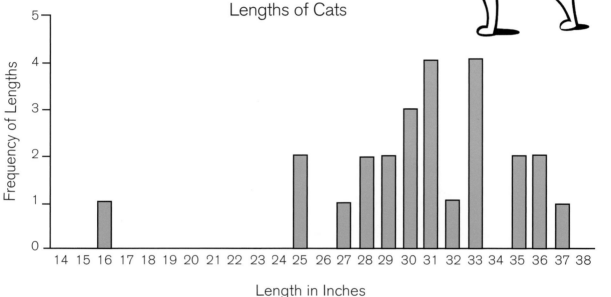

1. How many cats measured 30 inches long from nose to tail?_____ How can you tell?_____
   _____

2. How many cats were measured in all?_____ How can you tell? _____
   _____

3. If you added the lengths of the three shortest cats, what would the total of those lengths be? _____
   How can you tell? _____
   _____

4. What is the typical length of a cat from nose to tail?_____ Explain why you think your
   answer is correct._____
   _____

5. If we measured another cat, how long do you think it would be? _____ Explain why you think your
   answer is correct. _____
   _____

TV Watching

Name _____

The forty-nine students in two seventh-grade classes were asked to report the number of hours they watched TV the previous week. Here are their data, listed in order from least to greatest number of hours.

Number of Hours of TV Watching in One Week

0, 0, 0, 1, 3, 3, 4, 4, 5, 5, 5, 5, 6, 6, 6, 6, 6, 6, 6, 6, 7, 7, 7, 7, 7, 7, 8, 8, 8, 8, 8, 9, 11, 11, 13, 13, 13, 13, 14, 14, 14, 15, 15, 20, 20, 20, 22, 22, 30

1. On grid paper, make a graph of the data that might help the parents of these students decide whether the students are watching too much TV.

2. About how many hours of TV did the majority of students watch per day? _____
   About how many hours per day is represented by the greatest value? _____
   By the least value? _____

3. Do you think these seventh-grade students spend too much time watching TV? _____
   Do you think that if their parents read your graph, they would conclude that these students spend too much time watching TV? _____ Explain how you reached your conclusions. _____

   _____

   _____

   _____

   _____

   _____

Making the Data

Name _____

1. Karl has eight people in his family. He wondered how many hours of TV each of them might have watched in a week if the mean of the eight values were five hours. Write down eight amounts of TV-watching time that have a mean of five hours. _____
    Is there more than one set of eight amounts that has a mean of five hours? _____

2. When Karl told his father what he had done, his father wondered how the values might change if the mean were five hours and the median were four hours. Write down eight amounts of TV-watching time that have a mean of five hours *and* a median of four hours. _____
    Is there more than one data set that satisfies both conditions? _____

3. Karl's mother challenged him to write down eight values with a mean of five hours, a median of four hours, and a range of seven hours. Write down eight possible amounts of TV-watching time that have a mean of five hours, a median of four hours, *and* a range of seven hours. _____
    _____Is there more than one data set that satisfies all the conditions? _____

4. List one of the strategies your classmates used to solve problem 3 that was different from the strategy you used. _____

    _____

    Using the strategy you described, solve problem 3 again._____

    _____

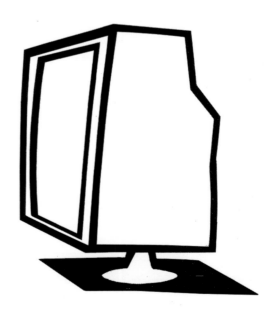

Drop Off

Name _____

The graph below shows the greatest drop in fifty-five major roller coasters in the United States. These data were reported in whole numbers of feet.

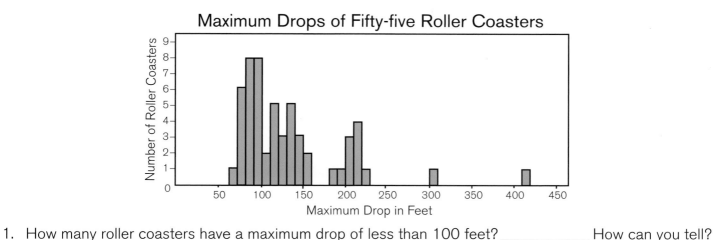

1. How many roller coasters have a maximum drop of less than 100 feet? _____ How can you tell?
   _____
   _____

2. What does the gap between 160 and 180 tell you? _____
   _____
   _____
   _____

3. What do the two rightmost bars tell you? _____
   _____
   _____
   _____

4. How do the roller coasters represented by the small cluster compare with the roller coasters represented by the large cluster? _____
   _____
   _____

5. The mean and median of the individual data values are 115 and 132, *but not necessarily in that order.* Which value is the mean? _____ Which value is the median? _____ How do you know?
   _____
   _____
   _____

6. Which of those two values was easier for you to identify? _____
   Which was more difficult to identify? _____ Why? _____
   _____
   _____
   _____

Students and Basketball Players

Name _____

How much taller are basketball players than students? Jane and Sam had some data that showed the heights in centimeters of the students in their class and of twenty-five professional basketball players. Jane made stem plots to display these heights.

Heights of Students

```
13 | 8  8  8  9
14 | 1  2  7  7  7
15 | 0  0  1  1  1  2  2  2  2  3  3  6  6  7  8
16 |
17 | 1
```

Key: 14 | 7 means 147 cm

Heights of Basketball Players

```
18 | 0  3  5
19 | 0  2  5  7  8  8
20 | 0  0  2  3  5  5  5  5  7
21 | 0  0  0  1  4  5
22 | 0
```

Key: 19 | 2 means 192 cm

1. How many students are in the class? _____ How many heights of basketball players have been reported? _____ How can you tell? _____
   _____
   _____

2. How many students are 152 cm tall? _____ How can you tell? _____
   _____

3. How many basketball players are at least 198 cm tall? _____ How can you tell?_____
   _____

4. What is the typical height of the students? _____ Explain how you arrived at your answer. _____
   _____
   _____
   _____

5. What is the typical height of the basketball players? _____ Explain how you arrived at your answer.
   _____
   _____
   _____

6. About how much taller are the basketball players than the students in this class? _____
   Explain how you arrived at your answer._____
   _____
   _____

Batteries

Name _____

Mrs. Brewer uses graphing calculators in her mathematics classes. She wants to use longer-lasting batteries so that a minimum of time is lost because of dead batteries. She searched the Internet and found the following data on the life, in hours, of two brands of batteries.

Battery Life in Hours

Always Ready Batteries				Tough Cell Batteries		
96	111	110		101	95	103
115	95	115		91	121	106
106	86	110		104	121	111
115	63	84		84	110	114
44	110	52		103	94	107
110	111	107		121	97	99
115	116	89		107	109	96
113	92	93		97	105	110
74	117	101		103	112	110
115	175	101		115	115	98
114	31	105		140	102	108
19	105	99		130	91	110
86	22			86	95	
203	96			100	88	

1. On grid paper, make two graphical representations that illustrate the differences between these two brands of batteries.

2. What are the quartiles for each set of data? _____

   _____

   What do these values indicate about the two brands?_____

   _____

   _____

3. Which brand of batteries seems to have the longer life? _____
   Explain the reasons for your answer. _____

   _____

   _____

4. Suppose that Mrs. Brewer puts new batteries of the brand you named in question 3 in all the calculators at the beginning of the school year. About when during the year will she need to have replacements available? _____ Explain the reasons for your answer. _____

   _____

   _____

   _____

Stopping Distances

Name _____

Kelly is shopping for a new car. She is very concerned about safety. She found these data about braking distances for ten different cars of each of the two models that she is most seriously considering.

Braking Distances in Feet

Small Sedan	At 30 MPH	At 60 MPH
Car 1	65	155
Car 2	63	147
Car 3	68	146
Car 4	67	154
Car 5	65	151
Car 6	58	145
Car 7	67	125
Car 8	57	148
Car 9	66	147
Car 10	68	154

Large Sedan	At 30 MPH	At 60 MPH
Car 1	70	161
Car 2	53	130
Car 3	61	135
Car 4	64	160
Car 5	76	168
Car 6	55	137
Car 7	80	139
Car 8	54	133
Car 9	60	137
Car 10	52	138

1. Which model seems to have the shortest stopping distance at 30 MPH? _____

    Using the grid paper to make an appropriate representation of the data, explain your answer. _____

    _____

    _____

    _____

2. Which model seems to have the shortest stopping distance at 60 MPH? _____

    Using the grid paper to make an appropriate representation of the data, explain your answer _____

    _____

    _____

    _____

3. What conclusion can you draw from the data about the overall stopping distances of these two models?

    _____

    Using the grid paper to make an appropriate representation of the data, explain your answer _____

    _____

    _____

    What else would you like to know about the safety of these two models? _____

    _____

    _____

Classroom Climate

Name _____

At Madison Middle School, some classrooms always seem to be colder than others. The two-story school faces south, and each classroom has windows to the outside. Mary thinks the classrooms on the north side of the building are colder than those on the south side, and Diane thinks the classrooms on the first floor are colder than those on the second floor.

The students decide to record the temperatures at four times of the day in four classrooms:

- The mathematics classroom on the second floor
- The English classroom on the first floor directly below the mathematics room
- The science classroom across from the mathematics classroom
- The art classroom directly below the science room

The windows in the mathematics and English classrooms face south, whereas the windows in the other two classrooms face north.

The students set up a calculator-based laboratory (CBL) with a temperature probe in each of the four classrooms. They decide to analyze the 6 A.M., 9 A.M., noon, and 3 P.M. temperature readings for each day. The data, recorded in degrees Fahrenheit, are given in the chart below:

Classroom Temperatures

	Time	Temperature (°F) in Mathematics, Room 203, South Facing	Temperature (°F) in English, Room 103, South Facing	Temperature (°F) in Science, Room 204, North Facing	Temperature (°F) in Art, Room 104, North Facing
Day 1	6 A.M.	63	62	62	60
	9 A.M.	66	65	65	65
	Noon	69	67	68	68
	3 P.M.	71	72	71	70
Day 2	6 A.M.	64	65	63	62
	9 A.M.	67	67	67	66
	Noon	70	70	69	70
	3 P.M.	74	73	70	71
Day 3	6 A.M.	64	62	61	60
	9 A.M.	68	68	67	65
	Noon	72	71	70	71
	3 P.M.	74	72	72	72
Day 4	6 A.M.	66	63	64	61
	9 A.M.	70	68	68	66
	Noon	75	70	73	71
	3 P.M.	74	71	73	70
Day 5	6 A.M.	66	65	64	62
	9 A.M.	69	68	67	67
	Noon	74	73	71	71
	3 P.M.	73	74	72	69

1. Are the classrooms on the first floor colder than those on the second floor? _____
   If so, by about how much? _____

2. Are the classrooms facing north colder than those facing south? _____
   If so, by about how much? _____

Cereal

Name _____

Ms. Chan's class went to a local grocery store and recorded the amount of sugar per serving of the breakfast cereals that were on sale. The graph below shows their results.

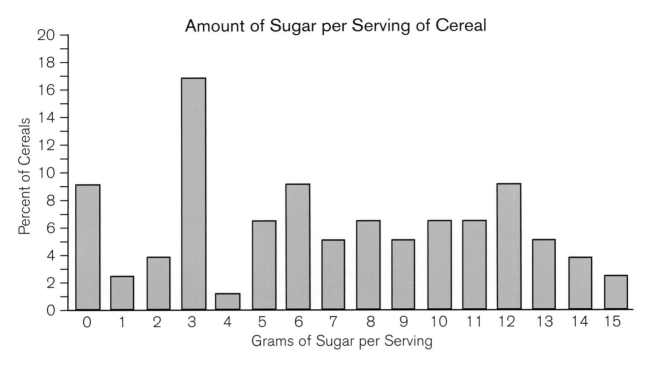

Amount of Sugar per Serving of Cereal

1. What percent of the cereals had three grams or less of sugar per serving?_____

2. What percent of the cereals had more than nine grams of sugar per serving?_____

3. How many different cereals did the students gather information about?_____

_____

_____

4. Did more cereals have less than eight grams of sugar per serving, or did more cereals have eight grams or more of sugar per serving? _____

5. What is the median amount of sugar? _____

6. What are the quartiles?_____

_____

_____

_____

_____

_____

A Matter of Opinion

Name _____

The new middle school is choosing a mascot. The possibilities have been narrowed to two: Hornets and Panthers. Latisha and Julio decided to survey the community to gauge reactions to these choices, so on Saturday they went to the mall. Latisha stood in front of the east entrance for two hours, and Julio stood in front of the west entrance for four hours. Each of them asked passersby to respond to a question. Thirty people responded to Latisha's survey, and sixty people responded to Julio's survey. Here are the questions they asked and the numbers of each response they received.

Latisha's question and data:

Which do you think is the better name for our mascot—Hornets or Panthers?
Hornets: 15 people
Panthers: 10 people
Don't care: 5 people

Julio's question and data:

Do you like Hornets better than Panthers as the name of our mascot?
Yes: 24 people
No: 20 people
Don't care: 16 people

1. Do you think the wording of the question influenced the responses? _____ Explain. _____
_____
_____

2. On grid paper, make a double-bar graph for these two data sets.

3. From the graph, which question seems to have generated more support for Hornets? _____
_____
_____

4. From the graph, which question seems to have generated more support for Panthers? _____
_____
_____

5. From the graph, which question seems to have generated a higher level of no opinion? _____
_____
_____

6. Make a relative-frequency double-bar graph for the two data sets.

7. How is this graph different from the graph you made for task 2? _____
_____
_____

8. If you wanted to argue in support of the name Hornets, which data would you use? _____
Explain your answer. _____
_____
_____

9. If you wanted to argue in support of the name Panthers, which data would you use? _____
Explain your answer. _____
_____
_____

Migraines: Histograms

Name _____

Below are data collected from patients who suffer from migraine headaches. The patients were instructed to take their assigned drugs as soon as their headaches began and to record how much time passed before the drugs gave relief. Drug A is a traditional drug, and Drug B is an experimental drug. Each value is the number of minutes (rounded to the nearest two minutes) that elapsed before a patient got relief.

Drug A (106 patients)

16, 18, 18, 20, 22, 22, 24, 24, 26, 26, 28, 28, 30, 30, 32, 32, 34, 36, 36, 36, 38, 38, 40, 42, 44, 44, 46, 46, 48, 50, 54, 56, 56, 58, 58, 58, 62, 62, 64, 64, 66, 68, 68, 70, 70, 70, 72, 72, 74, 76, 76, 76, 78, 78, 80, 80, 80, 82, 82, 84, 84, 84, 86, 86, 88, 88, 88, 88, 90, 90, 90, 90, 90, 92, 92, 92, 92, 94, 94, 94, 96, 96, 98, 98, 98, 98, 100, 100, 100, 100, 102, 102, 102, 104, 104, 106, 106, 108, 108, 108, 110, 110, 112, 114, 118, 120

Drug B (47 patients)

18, 20, 20, 22, 24, 24, 24, 26, 26, 30, 30, 30, 34, 34, 34, 36, 36, 36, 38, 38, 40, 40, 44, 44, 46, 50, 52, 52, 56, 56, 58, 62, 62, 66, 74, 74, 78, 88, 94, 98, 98, 100, 104, 106, 110, 116, 120

1. From examining these data, which drug do you think gave faster relief from headache pain? _____
   Explain. _____

   _____

   _____

2. Construct a histogram for each data set on the axes below. Title your display, and specify an appropriate scale on the horizontal axis.

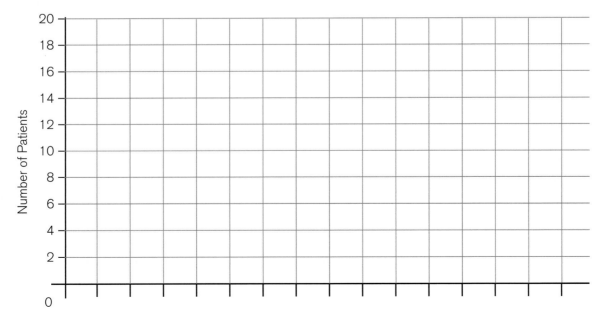

Time in Minutes to Effect

Migraines: Histograms (continued)

Name _____

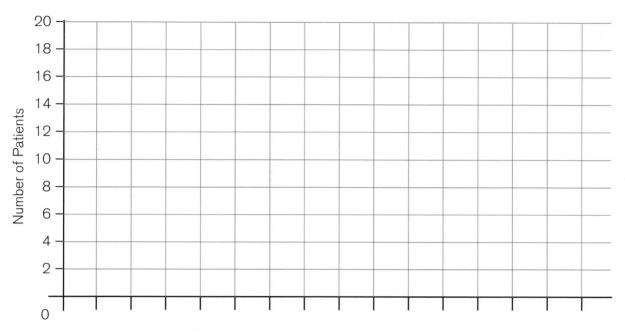

Time in Minutes to Effect

3. From examining the histograms, which drug do you think was more effective in giving fast relief from headache pain? _____ Explain. _____

_____

_____

4. Construct a relative-frequency histogram on the axes below.

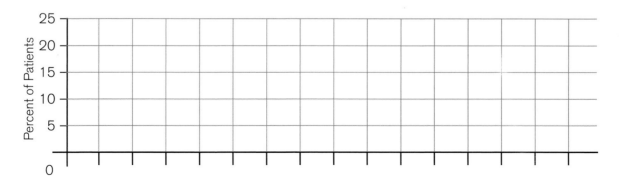

Time in Minutes to Effect

Name _____

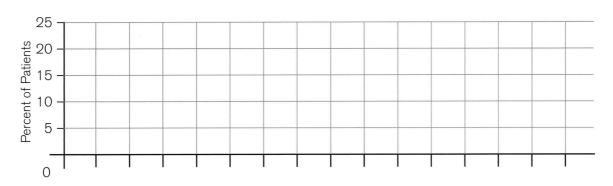

Time in Minutes to Effect

5. On the basis of your examination of the relative-frequency histograms, which drug do you think gave faster relief from headache pain? _____ Explain. _____

_____

_____

6. Some students get different answers for questions 3 and 5. Why do you think that happens?_____

_____

_____

7. How does changing the display change the information you can read from the graph?

_____

_____

_____

8. What advantage does the histogram have over the relative-frequency histogram?_____

_____

_____

9. What advantage does the relative-frequency histogram have over the histogram?_____

_____

_____

_____

Migraines: Box Plots

Name _____

The data below show the number of minutes that elapsed before patients, some taking drug A and some taking drug B, got relief from migraine headaches.

Drug A (106 patients)

16, 18, 18, 20, 22, 22, 24, 24, 26, 26, 28, 28, 30, 30, 32, 32, 34, 36, 36, 36, 38, 38, 40, 42, 44, 44, 46, 46, 48, 50, 54, 56, 56, 58, 58, 58, 62, 62, 64, 64, 66, 68, 68, 70, 70, 70, 72, 72, 74, 76, 76, 76, 78, 78, 80, 80, 80, 82, 82, 84, 84, 84, 86, 86, 88, 88, 88, 88, 90, 90, 90, 90, 90, 92, 92, 92, 92, 94, 94, 94, 96, 96, 98, 98, 98, 98, 100, 100, 100, 100, 102, 102, 102, 104, 104, 106, 106, 108, 108, 108, 110, 110, 112, 114, 118, 120

Drug B (47 patients)

18, 20, 20, 22, 24, 24, 24, 26, 26, 30, 30, 30, 34, 34, 34, 36, 36, 36, 38, 38, 40, 40, 44, 44, 46, 50, 52, 52, 56, 56, 58, 62, 62, 66, 74, 74, 78, 88, 94, 98, 98, 100, 104, 106, 110, 116, 120

One way to represent these data is with dot plots like those below.

Time to Effect of Drug A

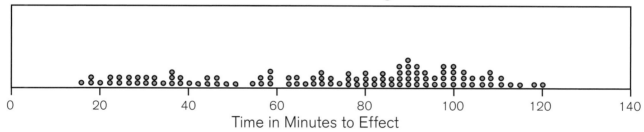

Time in Minutes to Effect

Time to Effect of Drug B

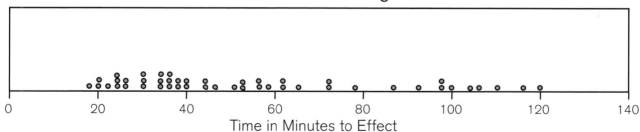

Time in Minutes to Effect

1. Use a vertical line segment that intersects the horizontal axis to mark the median of each data set.

2. What do you know about the number of data values on either side of the median? _____

_____

_____

Name _____

3. On the basis of a comparison of only the medians, which drug appears to provide faster relief? _____
   Explain your answer. _____
   _____
   _____

4. Use a vertical line segment that intersects the horizontal axis to mark the median of each half of each
   set of data. You should now have three vertical line segments drawn on the dot plot for each data set.
   The two new values that you have identified are called the *lower quartile* (or *first quartile*) and the
   upper quartile (or *third quartile*).

5. What do you know about the number of data elements in each of the four intervals for drug A?_____
   _____For drug B?_____

6. Look again at the lower quartiles, the medians, and the upper quartiles that you marked on the dot plots.
   On the basis of a comparison of these three values in the data sets, which drug appears to provide
   faster relief? _____ Explain your answer.
   _____
   _____

7. Use a vertical line segment to mark each extreme (least and greatest) value in each data set. You
   should now have five values marked in each data set. These five values are called the *five-point summary* of a data set.

8. Use the five-point summaries and the axis below to make two box plots. Label the axis and title your
   display.

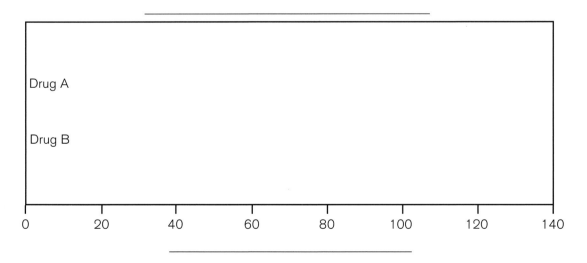

9. What can you say about the data for drug A in the first interval compared with the data for drug A in the
   fourth interval?_____
   _____

Migraines: Box Plots (continued)

Name ______________________________

10. What can you say about the data for drug B in the first interval compared with the data for drug B in the fourth interval? ______________________________

____

____

11. What does the distance between the vertical line segments tell you about how the data are spread out?

____

12. The lower quartile for drug A is about the same as the median for drug B. What does that information tell you about how the speeds of the drugs compare? ______________________________

____

____

13. The median for drug A is about the same as the upper quartile for drug B. What does that information tell you about how the speeds of the drugs compare? ______________________________

____

____

14. On the basis of all the information, which drug seems to provide faster relief? ______________

Explain your answer. ______________________________

____

____

Reading a Scatterplot

Name _____

The graph below shows the population and the number of area codes for each state in the United States. Use the scatterplot to answer the questions on the next page.

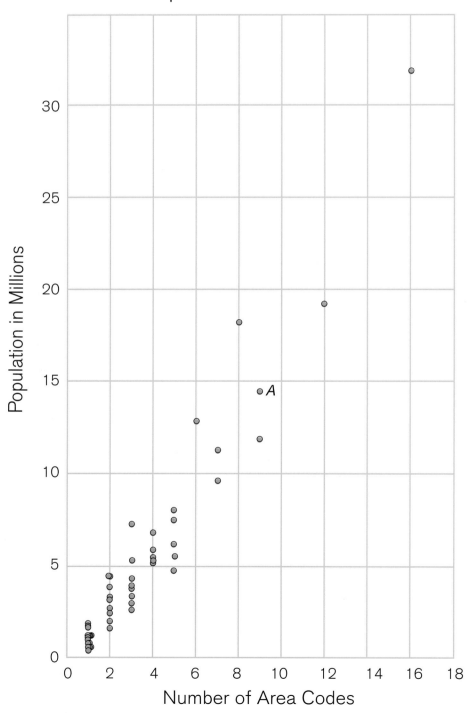

State Population and Area Codes

Reading a Scatterplot (continued)

Name _____

1. How many states have five area codes? _____ How did you determine the number?

   _____

   _____

2. What does the point labeled "A" represent? _____

   How did you decide what it represented? _____

   _____

3. The population of Canada is approximately 29,100,000. If a similar pattern exists there, how many area

   codes do you think Canada has? _____ Explain your answer._____

   _____

   _____

4. The population of Great Britain is approximately 58,600,000. If Great Britain used the same system,

   how many area codes do you think it would need? _____ Explain your answer. _____

   _____

   _____

5. Suppose that a state had fourteen area codes. What do you think the population of that state would be?

   _____ Explain your answer. _____

   _____

   _____

6. Describe the relationship between a state's population and the number of area codes assigned to it.

   _____

   _____

   _____

   _____

   _____

Congress and Pizza

Name _____

What do pizzas have to do with the United States House of Representatives? In particular, how does the number of pizza restaurants in a state relate to the number of U.S. representatives for that state?

Below is a table showing the number of pizza restaurants and U.S. representatives for forty states.

Congress and Pizza

State	Number of U.S. Representatives	Number of Pizza Restaurants	State	Number of U.S. Representatives	Number of Pizza Restaurants
Alabama	7	1071	Nevada	3	406
Alaska	1	149	New Hampshire	2	119
Arizona	8	919	New Jersey	13	1182
Arkansas	4	776	New Mexico	3	454
Colorado	7	929	North Carolina	13	1673
Connecticut	5	732	North Dakota	1	165
Delaware	1	133	Oklahoma	5	1034
Georgia	13	1515	Oregon	5	816
Hawaii	2	217	Pennsylvania	19	1682
Idaho	2	298	Rhode Island	2	136
Indiana	9	1394	South Carolina	6	1011
Iowa	5	838	South Dakota	1	158
Kansas	4	708	Tennessee	9	1652
Kentucky	6	879	Utah	3	469
Louisiana	7	952	Vermont	1	128
Maine	2	212	Virginia	11	1701
Maryland	8	721	Washington	9	1357
Missouri	9	1195	West Virginia	3	314
Montana	1	236	Wisconsin	8	1286
Nebraska	3	471	Wyoming	1	108

Sources of data: U.S. Department of Commerce, U.S. Census Bureau (n.d.a; n.d.b)

1. Represent the data to determine if a relationship exists between the number of pizza restaurants and the number of U.S. representatives. You may use grid paper for your representation.

2. Describe any relationship you see. _____

_____

_____

_____

People, Congress, and Pizza

Name _____

Below is a table showing the population, the number of U.S. representatives, and the number of pizza restaurants for forty states.

People, Congress, and Pizza

State	Population (2000)	Number of Representatives	Number of Pizza Restaurants	State	Population (2000)	Number of Representatives	Number of Pizza Restaurants
Alabama	4,461,130	7	1071	Nevada	2,002,032	3	406
Alaska	628,933	1	149	New Hampshire	1,238,415	2	119
Arizona	5,140,683	8	919	New Jersey	8,424,354	13	1182
Arkansas	2,679,733	4	776	New Mexico	1,823,821	3	454
Colorado	4,311,882	7	929	North Carolina	8,067,673	13	1673
Connecticut	3,409,535	5	732	North Dakota	643,756	1	165
Delaware	785,068	1	133	Oklahoma	3,458,819	5	1034
Georgia	8,206,975	13	1515	Oregon	3,428,543	5	816
Hawaii	1,216,642	2	217	Pennsylvania	12,300,670	19	1682
Idaho	1,297,274	2	298	Rhode Island	1,049,662	2	136
Indiana	6,090,782	9	1394	South Carolina	4,025,061	6	1011
Iowa	2,931,923	5	838	South Dakota	756,874	1	158
Kansas	2,693,824	4	708	Tennessee	5,700,037	9	1652
Kentucky	4,049,431	6	879	Utah	2,236,714	3	469
Louisiana	4,480,271	7	952	Vermont	609,890	1	128
Maine	1,277,731	2	212	Virginia	7,100,702	11	1701
Maryland	5,307,886	8	721	Washington	5,908,684	9	1357
Missouri	5,606,260	9	1195	West Virginia	1,813,077	3	314
Montana	905,316	1	236	Wisconsin	5,371,210	8	1286
Nebraska	1,715,369	3	471	Wyoming	495,304	1	108

Sources of data: U.S. Department of Commerce, U.S. Census Bureau (n.d.a; n.d.b)

1. Make a scatterplot to display the relationship between state population and the number of U.S. representatives. Then make a scatterplot to display the relationship between state population and the number of pizza restaurants.

2. Compare the two scatterplots with the scatterplot you made earlier that shows the relationship between the number of U.S. representatives and the number of pizza restaurants in each state. Which pair of variables seems to have the strongest relationship? Explain your answer. _____

_____

_____

3. How is state population related to the number of U.S. representatives and to the number of pizza restaurants? _____

Predicting

Name _____

The data in People, Congress, and Pizza were not complete; data for ten states were left out of the list. A partially completed table is given below.

People, Congress, and Pizza

State	Population (2000)	Number of Representatives	Number of Pizza Restaurants
California	_____	53	_____
Florida	_____	_____	2505
Illinois	_____	19	_____
Massachusetts	_____	_____	620
Michigan	_____	15	_____
Minnesota	_____	_____	933
Mississippi	_____	4	_____
New York	_____	_____	2574
Ohio	_____	18	_____
Texas	_____	_____	5200

Sources of data: U.S. Department of Commerce, U.S. Census Bureau (n.d.a; n.d.b)

1. Use two of your scatterplots for the number of U.S. representatives, the number of pizza restaurants, and the population to predict or interpolate the missing data. Explain how you made your predictions.

   _____

   _____

   _____

   _____

   _____

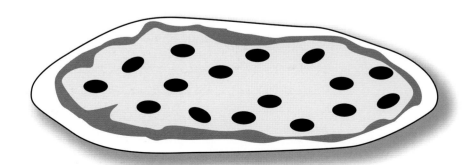

Predicting (continued)

Name _____

Below is the actual information for the ten states. Compare your results with the actual information.

People, Congress, and Pizza

State	Population (2000)	Number of Representatives	Number of Pizza Restaurants
California	33,930,798	53	6524
Florida	16,028,890	25	2505
Illinois	12,439,042	19	3047
Massachusetts	6,355,568	10	620
Michigan	9,955,829	15	1838
Minnesota	4,925,670	8	933
Mississippi	2,852,927	4	685
New York	19,004,973	29	2574
Ohio	11,374,540	18	2387
Texas	20,903,994	32	5200

Sources of data: U.S. Department of Commerce, U.S. Census Bureau (n.d.a; n.d.b)

2. How well did the scatterplots allow you to predict or interpolate the information? _____

_____

_____

_____

_____

3. Which predicted or interpolated values were close to the actual values? _____

Which values were far from the actual ones? _____

What could account for the large differences in the values that were far off? _____

_____

_____

_____

_____

Line Segments for Approximating Lines of Fit

Copy this master onto transparencies, cut on the dashed lines, and give several line segments to each student to use to approximate lines of fit on a scatterplot.

Olympic Gold Times

Name _____

Below are bivariate data for the Olympic winning times in the men's 200-meter dash.

Olympic Gold-Medal Times for the Men's 200-Meter Dash

Year	Time in Seconds	Year	Time in Seconds
1900	22.2	1956	20.6
1904	21.6	1960	20.5
1908	22.6	1964	20.3
1912	21.7	1968	19.8
1920	22.0	1972	20.0
1924	21.6	1976	20.23
1928	21.8	1980	20.19
1932	21.2	1984	19.80
1936	20.7	1988	19.75
1948	21.1	1992	20.01
1952	20.7	1996	19.32

Source of data: Brunner (1999)

1. Is there a relationship between the year and the winning time? (*Hint:* Make a scatterplot.) _____
   If so, describe it. _____
   _____
   _____

2. Interpolate or predict the winning times for the following years.

Year	Winning Time in Seconds
1916	_____
1940	_____
1944	_____
2000	_____
2004	_____
2008	_____
2012	_____
2016	_____

3. Using an almanac or other data source (e.g., the Internet), compare your predicted winning time for the year 2000 with the actual winning time for the event in that year. _____
   _____
   _____

4. What are the limitations, if any, of predicting the winning times for the years 2012 and 2016? How far into the future are predictions for winning times likely to be reasonable? _____
   Explain your answer. _____
   _____
   _____

Population Trends

Name _____

Below are data about the population of the United States in each census year.

United States Population 1790–2000

Year	Population to the Nearest 100,000	Year	Population to the Nearest 100,000
1790	3,900,000	1900	76,000,000
1800	5,300,000	1910	92,000,000
1810	7,200,000	1920	105,700,000
1820	9,600,000	1930	122,800,000
1830	12,900,000	1940	131,700,000
1840	17,000,000	1950	150,700,000
1850	23,200,000	1960	179,300,000
1860	31,400,000	1970	203,200,000
1870	39,800,000	1980	226,500,000
1880	50,200,000	1990	248,700,000
1890	62,900,000	2000	281,400,000

Source of data: U.S. Department of Commerce, U.S. Census Bureau (n.d.c)

1. Create a display of the data, and determine the relationship between the year and the population of the United States. Describe the relationship. _____

_____

_____

2. Use your data display to predict the population of the United States for the next five decades.

Year	Population (to the Nearest 100,000)
2010	_____
2020	_____
2030	_____
2040	_____
2050	_____

3. How accurate do you think your predictions are? _____ What factors might limit the accuracy of your predictions? _____

_____

_____

What other information might help you make better predictions?_____

_____

_____

_____

Time Is Money?

Name _____

Below are data about the running times and the gross receipts for thirty of the top movies of 1997.

Running Times and Gross Receipts of Movies

Title	Running Time in Minutes	Gross Receipts in Millions of Dollars	Title	Running Time in Minutes	Gross Receipts in Millions of Dollars
Air Force One	124	173	In and Out	92	64
Anaconda	90	66	The Jackal	124	55
Anastasia	90	58	Jungle 2 Jungle	105	60
As Good As It Gets	130	148	Kiss the Girls	117	60
Batman and Robin	125	107	L. A. Confidential	136	65
Con Air	105	101	Liar, Liar	87	181
Conspiracy Theory	135	76	The Lost World:		
Contact	150	101	Jurassic Park	129	229
Dante's Peak	112	67	Men in Black	98	250
The Devil's Advocate	144	61	Mouse Hunt	97	62
Face Off	140	112	My Best Friend's		
The Fifth Element	125	64	Wedding	105	127
Flubber	93	93	The Saint	118	61
George of the Jungle	91	105	Scream 2	120	101
Good Will Hunting	126	138	Tomorrow Never Dies	119	125
Hercules	92	99			
I Know What You Did Last Summer	100	72			

Sources of data: "Top Movies, 1997" (n.d.) and Craddock (2001)

1. Is there a relationship between the running time and the gross receipts? _____
   If so, describe the relationship. _____
   _____
   _____

2. What factors might influence the gross receipts of a film?_____
   _____
   _____
   _____

Solutions for the Blackline Masters

Solutions for "Lengths of Cats"

1. Three; the bar over 30 has height 3, indicating three cats.
2. Twenty-five; add the heights of all the bars to determine the total frequency.
3. Sixty-six inches; the shortest cat measures sixteen inches, and the next two each measure twenty-five inches. $16 + 25 + 25 = 66$.
4. Typical answers are twenty-seven to thirty-three inches, twenty-eight to thirty-one inches, and thirty to thirty-three inches.
5. Students may repeat the answer given to problem 4. Other answers include thirty-one or thirty-three inches (the modes) or thirty-one inches (the median).

Solutions for "TV Watching"

1. The students will typically construct either a line plot (see the example in fig. 1.3), a bar graph (see the example in fig. 1.6), or a stem plot.
2. Five to eight hours or a similar interval is acceptable; thirty hours per week is an average of approximately 4.3 hours per day; zero hours per week is necessarily zero hours per day.
3. Answers will vary. See the discussion section of the activity.

Solutions for "Making the Data"

1. One possibility for eight TV-watching amounts that have a mean of five hours is 2, 3, 4, 5, 5, 6, 7, 8. Many other data sets are possible.
2. Possibilities for eight TV-watching amounts that have a mean of five hours and a median of four hours are 2, 3, 4, 4, 4, 7, 8, 8 or 2, 3, 3, 3, 5, 8, 8, 8. Many other data sets are possible.
3. One possibility for eight TV-watching amounts that have a mean of five hours, a median of four hours, and a range of seven hours is 2, 2, 4, 4, 4, 7, 8, 9. Other data sets are possible.
4. Answers will vary.

Solutions for "Drop Off"

1. Twenty-three; adding the heights of the bars over the intervals of values less than 100 gives $1 + 6 + 8 + 8 = 23$.
2. The gap indicates that for some reason not revealed by the data, no roller coasters in the data set have maximum drops of 160 to 180 feet.
3. The two rightmost bars represent one roller coaster with a maximum drop of 300 to 310 feet and one with a maximum drop of 410 to 420 feet. Those two coasters would be very scary.
4. The roller coasters represented by the small cluster have greater maximum drops than those represented by the large cluster.
5. The median is 115; the mean is 132. The median is the middle value in a data set and the twenty-eighth data value is in the bar over the interval 110–120, so it must be 115. The right part of the distribution has two extreme values, which make the mean greater than the median, so the mean must be 132.
6. Answers will vary.

Solutions for "Students and Basketball Players"

1. Twenty-five students are in the class; twenty-five basketball players' heights have been reported. The answers are found by counting the number of leaves in each stem plot.
2. Four students are 152 cm tall. There are four leaves with the value 2 to the right of the 15 stem.

3. Eighteen basketball players are at least 198 cm tall. The answer can be determined by counting the number of leaves representing a value of 198 cm or greater.

4. One possible answer is the median, 151 cm. Another is an interval, such as 150 to 153 cm, that captures a cluster of data.

5. One possible answer is the median, 203 cm. Another is an interval, such as 200 to 205 cm, that captures a cluster of data.

6. The basketball players are about 50 cm taller than the students. To obtain the solution, subtract the medians: $203 - 151 = 52$ cm, or about 50 cm.

Solutions for "Batteries"

1. Several graphs are acceptable. See the discussion section of the activity.

2. The quartiles for Always Ready batteries are 87.5, 105, and 113.5 hours; the quartiles for Tough Cell batteries are 97, 104.5, and 110.5 hours. The quartiles for Tough Cell batteries are closer together, so that brand is more consistent than Always Ready batteries.

3. The students may have different opinions. See the discussion section of the activity for possible reasons.

4. The median life of the two brands is close: 105.0 hours for Always Ready batteries and 104.5 hours for Tough Cell batteries. If a calculator is used for three hours each day, five days a week, the batteries would need to be replaced about every seven weeks. Other assumptions about the average use each day would lead to different conclusions.

Solutions for "Stopping Distances"

1. Answers will vary. The small sedan has more consistent stopping distances, but the large sedan has a cluster of data that represents shorter distances than the main cluster for the small sedan.

2. Answers will vary. The small sedan has more consistent stopping distances, but the large sedan has a cluster of data that represents shorter distances than the main cluster for the small sedan.

3. Answers will vary; however, each student's answer should be consistent with his or her answers to questions 1 and 2.

Solutions for "Classroom Climate"

1. Yes. The classrooms on the first floor are colder than those on the second floor by about 1°F. An explanation of how students might arrive at the solution is found in the discussion section of this activity.

2. Yes. The classrooms that face north are colder than those that face south by about 1°F. An explanation of how students might arrive at the solution is found in the discussion section of this activity.

Solutions for "Cereal"

1. 32.5 percent. (Students' answers may not be exact.)

2. 33.8 percent. (Students' answers may not be exact.)

3. It is impossible to tell from the graph.

4. More had less than eight grams of sugar per serving: approximately 55 percent had less than eight grams per serving, and approximately 45 percent had eight grams or more per serving.

5. The median is seven grams of sugar per serving. It is the amount of sugar per serving for which the cumulative percent is 50 percent.

6. First, or lower, quartile: three grams per serving; second quartile, or median: seven grams per serving; third, or upper, quartile: eleven grams per serving.

Solutions for "A Matter of Opinion"

1. Yes; Julio's question seems to direct responses toward Hornets.
2. This graph is displayed in figure 3.3.
3. Julio's bars are all taller than Latisha's bars, so his question *seems* to have generated more support, but he interviewed more people.
4. Julio's bars are all taller than Latisha's bars, so his question *seems* to have generated more support, but he interviewed more people.
5. Julio's bars are all taller than Latisha's bars, so his question *seems* to have generated more support, but he interviewed more people.
6. This graph is displayed in figure 3.4.
7. The relative heights of the bars are different.
8. Latisha's question seems to have generated the greater support for Hornets, so we would use her data.
9. The relative strength of support for Panthers seems to be the same for both data sets, so the argument would be supported equally well by either data set.

Solutions for "Migraines: Histograms"

1. It is difficult to tell from examining only the lists.
2. See the two histograms.

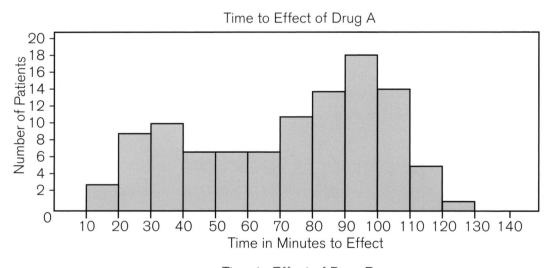

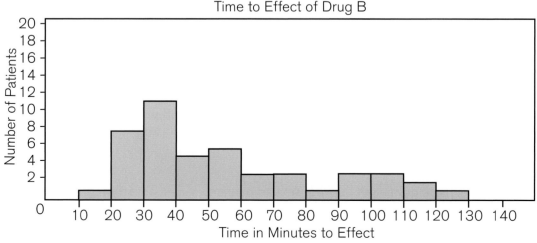

3. The bars for drug A are taller, so more people (although not a greater percent of people) got relief faster from drug A.

4. See the two relative-frequency histograms.

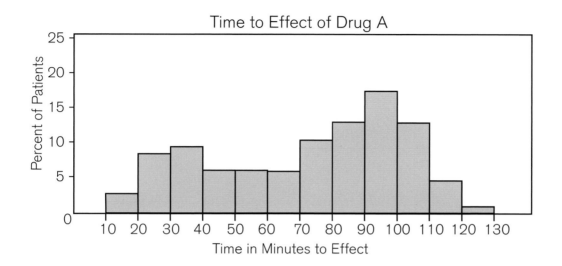

Time to Effect of Drug A

Vertical axis: Percent of Patients; Horizontal axis: Time in Minutes to Effect (10–130)

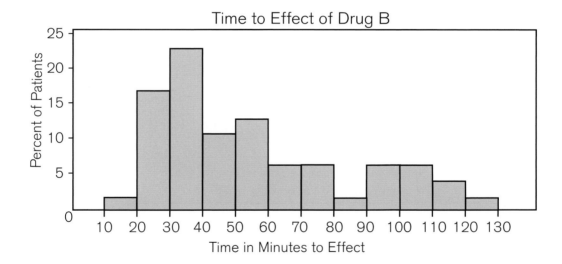

Time to Effect of Drug B

Vertical axis: Percent of Patients; Horizontal axis: Time in Minutes to Effect (10–130)

5. Drug B is faster; a greater percent of patients got faster relief.

6. The relative heights of the bars are different, so the visual impressions are different.

7. The vertical axis determines whether the bars represent counts of data values (in an absolute-frequency graph) or percents of data values (in a relative-frequency graph).

8. Actual counts can be read from a histogram.

9. The relative speeds of the drugs can be read from a relative-frequency histogram. In comparing the two data sets with unequal numbers of data values, it is more important to determine the percents of patients who received fast relief than to determine the actual numbers of patients.

Solutions for "Migraines: Box Plots"

1. See the middle line segment drawn on each of the dot plots.

Time to Effect of Drug A

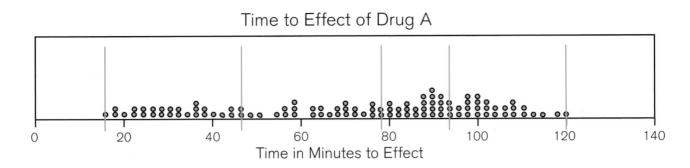

Time in Minutes to Effect

Time to Effect of Drug B

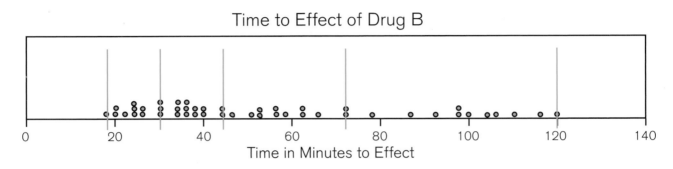

Time in Minutes to Effect

2. For each data set, the number of data values is the same on either side of the median.
3. Drug B; it appears to have provided relief to half the patients faster than drug A did.
4. See the second and fourth line segments drawn on the dot plots.
5. For each data set, the number of elements in each of the four intervals is the same.
6. Drug B; it appears to have provided relief to three-quarters of the patients in slightly less time than drug A provided relief to half the patients.
7. See the leftmost and rightmost line segments drawn on the dot plots.
8. See the box plots.

Time to Effect of Drugs A and B

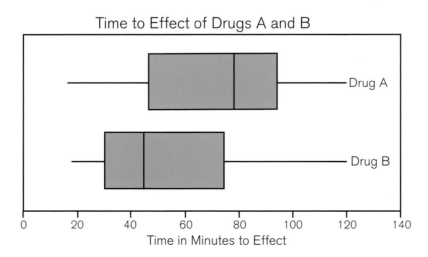

Time in Minutes to Effect

9. The number of values is the same.
10. The number of values is the same.

Navigating through Data Analysis in Grades 6–8

11. The closer the vertical line segments are, the more clustered the data are. The farther apart the vertical line segments are, the more spread out the data are.

12. Drug B gave faster relief to a greater fraction of patients.

13. Drug B gave faster relief to a greater fraction of patients.

14. Drug B gave faster relief to a greater fraction of patients.

Solutions for "Reading a Scatterplot"

1. Five states; count the number of dots above 5 on the horizontal (area code) axis.

2. Point A represents 14.5 million people and nine area codes; these values are determined by reading the *y*-coordinate and the *x*-coordinate.

3. About sixteen area codes; imagine a line of fit for these data, and then find the horizontal coordinate of the point associated with 29.1 million people.

4. About thirty-five area codes; a state with 20 million people would have approximately twelve area codes, and 58.6 million is a little less than three times 20 million, so Great Britain would have a bit fewer than three times twelve area codes, or approximately thirty-five area codes.

5. About 22 million people; imagine a line of fit for these data, and then find the *y*-coordinate of the point associated with fourteen area codes.

6. As the population increases, the number of area codes increases; the relationship is nearly linear.

Solutions for "Congress and Pizza"

1. See the scatterplot, one possible representation.

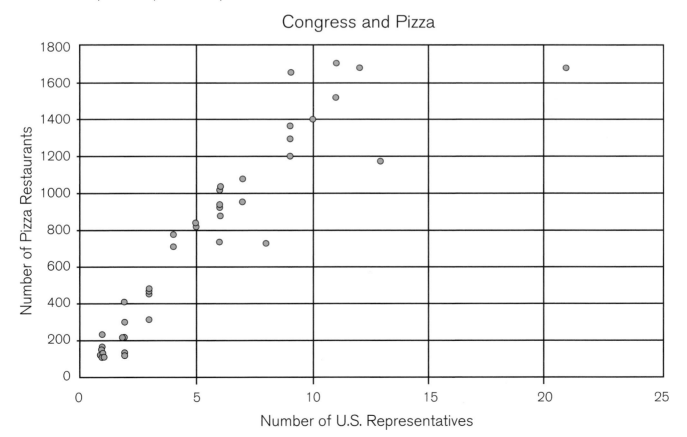

Congress and Pizza

2. As the number of representatives increases, so does the number of pizza restaurants. The increase is linear.

Solutions for "People, Congress, and Pizza"

1. See the scatterplots "State Population and Number of U.S. Representatives" and "State Population and Number of Pizza Restaurants." The scatterplot that the students made earlier, "Congress and Pizza," is also shown.

2. All three displays show an increasing linear relationship. The relationship between the number of representatives and the population of the states appears to be the strongest, since the points for that display look more like a line than the points in the other displays.

3. The number of representatives and the number of pizza restaurants increase as the population increases. In each example, the relationship appears to be linear.

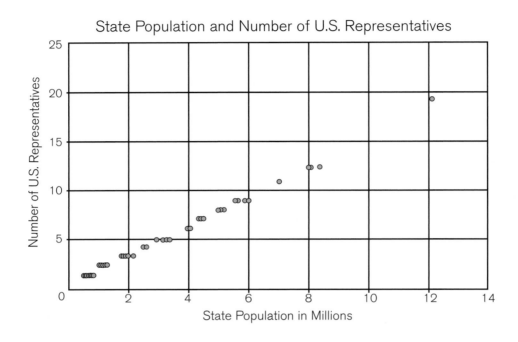

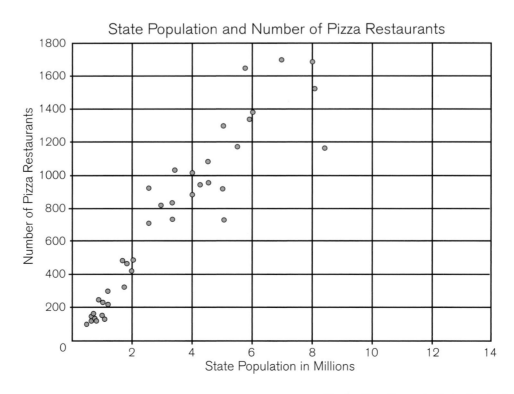

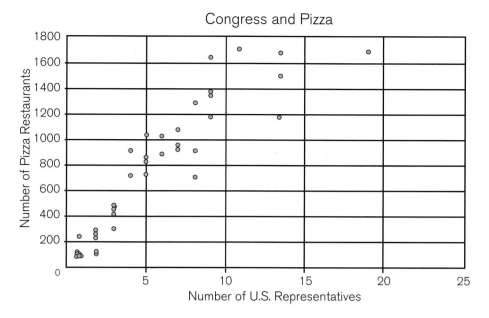

Congress and Pizza

Solutions for "Predicting"

1. The students' answers will vary, depending on the particular approximate line of fit used.
2. Answers will vary. See the "Discussion" section of the activity.
3. Answers will vary, depending on the particular approximate line of fit used. The close values are likely to be within the range of the data, and the predictions that are far from the actual values are likely to be outside the range of the data.

Solutions for "Olympic Gold Times"

1. Yes; the winning time for the race gets shorter as the year increases; that is, it shows a negative linear relationship. See the scatterplot.

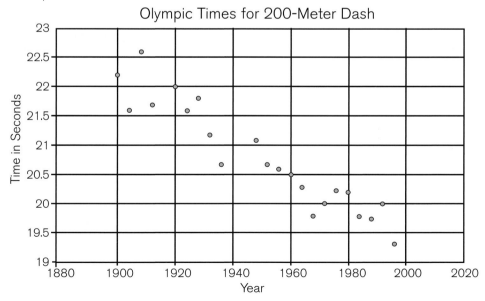

Olympic Times for 200-Meter Dash

2. Students' answers will vary. The values should decrease across the years.
3. Answers will vary.
4. There is a limit to how fast an athlete can run the 200-meter dash. At some point in the future, the winning time will "level off" at this barrier.

Solutions for "Population Trends"

1. See the graph. The population increases as the year increases. However, the relationship is not linear. The population is increasing faster in later years than in earlier years. Prior to 1950, the changes may reflect territorial acquisitions and the admission of new states as well as births and immigration.

2. Students' answers will vary.

3. The predictions for 2010 and 2020 might be more accurate than the predictions for the other three decades. However, many factors—for instance, baby booms, epidemics, the amount of livable land, and so on—can influence the population. Information about birth rates and mortality rates can help improve predictions.

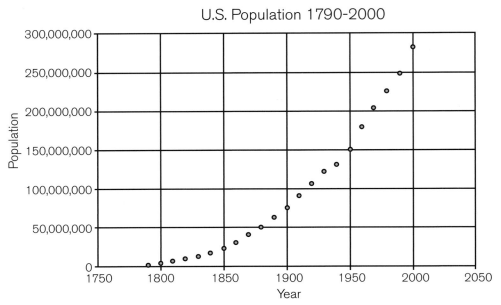

U.S. Population 1790–2000

Solutions for "Time Is Money?"

1. See the scatterplot. There appears to be no relationship between the running times and the gross receipts.

2. Possible factors are the type of movie, the popularity of particular actors in the movie, and the number of theaters showing it. More theaters can mean more receipts for the movie.

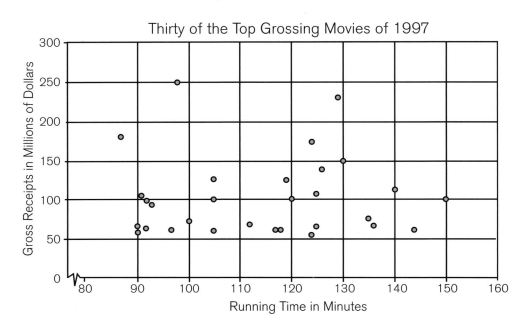

Thirty of the Top Grossing Movies of 1997

References

Brunner, Borgna. *Time Almanac 2000*. Boston, Mass.: Family Education Co., 1999.

Chapin, Suzanne, Alice Koziol, Jennifer MacPherson, and Carol Rezba. *Navigating through Data Analysis and Probability in Grades 3–5*. *Principles and Standards for School Mathematics* Navigations series. Reston, Va.: National Council of Teachers of Mathematics, 2002.

Craddock, Jim., ed. *Videohound's Golden Movie Retriever, 2001*. Farmington Hills, Mich.: Visible Ink Press, 2001.

Curcio, Frances R. "Comprehension of Mathematical Relationships Expressed in Graphs." *Journal for Research in Mathematics Education* 18 (November 1987): 382–93.

———. *Developing Data-Graph Comprehension in Grades K–8*. 2nd ed. Reston, Va.: National Council of Teachers of Mathematics, 2001.

Dixon, Juli K., and Christy J. Falba. "Graphing in the Information Age: Using Data from the World Wide Web." *Mathematics Teaching in the Middle School* 2 (March–April 1997): 298–304.

Friel, Susan N. "Teaching Statistics: What's Average?" In *The Teaching and Learning of Algorithms in School Mathematics*, 1998 Yearbook of the National Council of Teachers of Mathematics, edited by Lorna J. Morrow, pp. 208–17. Reston, Va.: National Council of Teachers of Mathematics, 1998.

Friel, Susan N., George W. Bright, and Frances R. Curcio. "Understanding Students' Understanding of Graphs." *Mathematics Teaching in the Middle School* 3 (November–December 1997): 224–27.

Friel, Susan N., Frances R. Curcio, and George W. Bright. "Making Sense of Graphs: Critical Factors Influencing Comprehension and Instructional Implications." *Journal for Research in Mathematics Education* 32 (March 2001): 124–58.

Friel, Susan N., and William T. O'Connor. "Sticks to the Roof of Your Mouth?" *Mathematics Teaching in the Middle School* 4 (March 1999): 404–11.

Graham, A. *Statistical Investigations in the Secondary School*. Cambridge: Cambridge University Press, 1987.

House, Peggy A. "Stand Up and Be Counted: The Mathematics of Congressional Apportionment." *Mathematics Teacher* 94 (November 2001): 692–97.

Kader, Gary D. "Means and MADs." *Mathematics Teaching in the Middle School* 4 (March 1999): 398–403.

Kader, Gary, and Mike Perry. "Learning Statistics with Technology." *Mathematics Teaching in the Middle School* 1 (September–October 1994): 130–36.

Kader, Gary D., and Mike Perry. "Pennies from Heaven—Nickels from Where?" *Mathematics Teaching in the Middle School* 3 (November–December 1997): 240–48.

Landwehr, James M., and Ann E. Watkins. *Exploring Data*. Palo Alto, Calif.: Dale Seymour Publications, 1986.

Matsumoto, Annette N. "Correlation, Junior Varsity Style." In *Teaching Statistics and Probability*, 1981 Yearbook of the National Council of Teachers of Mathematics, edited by Albert P. Shulte and James R. Smart, pp. 126–34. Reston, Va.: National Council of Teachers of Mathematics, 1981.

McClain, Kay. "Reflecting on Students' Understanding of Data." *Mathematics Teaching in the Middle School* 4 (March 1999): 374–80.

McClain, Kay, Paul Cobb, and Koeno Gravemeijer. "Supporting Students' Ways of Reasoning about Data." In *Learning Mathematics for a New Century*, 2000 Yearbook of the National Council of Teachers of Mathematics, edited by Maurice J. Burke, pp. 174–87. Reston, Va.: National Council of Teachers of Mathematics, 2000.

Mooney, Edward S. "A Framework for Characterizing Middle School Students' Statistical Thinking." *Mathematical Thinking and Learning* 4, no. 1 (2002): 23–63.

Moore, David S. *Statistics: Concepts and Controversies.* 3rd ed. New York: W. H. Freeman, 1991.

National Council of Teachers of Mathematics (NCTM). *Curriculum and Evaluation Standards for School Mathematics.* Reston, Va.: NCTM, 1989.

———. *Principles and Standards for School Mathematics.* Reston, Va.: NCTM, 2000.

North Carolina School of Science and Mathematics (NCSSM). Department of Mathematics and Computer Science. *Data Analysis.* New Topics for Secondary School Mathematics. Reston, Va.: National Council of Teachers of Mathematics, 1988.

Rubink, William L., and Sylvia R. Taube. "Mathematical Connections from Biology: 'Killer' Bees Come to Life in the Classroom." *Mathematics Teaching in the Middle School* 4 (March 1999): 350–56.

Scheaffer, Richard L. "Statistics for a New Century." In *Learning Mathematics for a New Century*, 2000 Yearbook of the National Council of Teachers of Mathematics, edited by Maurice J. Burke, pp. 158–73. Reston, Va.: National Council of Teachers of Mathematics, 2000.

"Top Movies, 1997." *Craig's Flick Picks.* Retrieved 28 July 2002 from www.craigbe.com.

U.S. Department of Commerce, U.S. Census Bureau. Economic Census, Sector 72: Accommodations and Foodservices, Miscellaneous Subjects, Principal Menu Type or Specialty, 1997. n.d.a. (Retrieved 25 July 2002 from www.census.gov.)

———. Table 1: Apportionment Population and Number of Representatives, by State: Census 2000. n.d.b. (Retrieved 1 December 2002 from www.census.gov/population/cen2000/tab01.txt.)

———. Table 2: Population, Housing Units, Area Measurements, and Density, 1790 to 1990. n.d.c. (Retrieved 25 July 2002 from www.census.gov /population/censusdata/table-2.pdf.)

VanLeuvan, Patricia. "Young Women Experience Mathematics at Work in the Health Professions." *Mathematics Teaching in the Middle School* 3 (November–December 1997): 198–206.

Wilson, Melvin R. (Skip), and Carol M. Krapfl. "Exploring Mean, Median, and Mode with a Spreadsheet." *Mathematics Teaching in the Middle School* 1 (September–October 1995): 490–95.

Zawojewski, Judith S., with Gary Brooks, Lynn Dinkelkamp, Eunice D. Goldberg, Howard Goldberg, Arthur Hyde, Tess Jackson, Marsha Landau, Hope Martin, Jeri Nowakowski, Sandy Paull, Albert P. Shulte, Philip Wagreich, and Barbara Wilmot. *Dealing with Data and Chance. Curriculum and Evaluation Standards for School Mathematics* Addenda Series, Grades 5–8. Reston, Va.: National Council of Teachers of Mathematics, 1991.